To Mom and Dad – because you always believed.

The Ecological Engineer
Volume One: KEEN Engineering

An Ecotone LLC Publication

For more information write:

Ecotone LLC
P.O. Box 7147
Kansas City, MO
64113-0147

Authors: David R. Macaulay, Jason F. McLennan
Book Design: Erin Gehle
Edited by: Fred McLennan

Library of Congress Control Number: 2005929646
Library of Congress Cataloging-in Publication Data

Macaulay, David R. and McLennan, Jason F.

The Ecological Engineer: Volume One: KEEN Engineering

ISBN 0-9749033-4-5

1. Engineering 2. Environment 3. Architecture

First Edition, Volume One

Printed in Canada on Post-Consumer Recycled Content Paper

Table of Contents

The Ecological Engineer
Volume One: KEEN Engineering

Acknowledgements

The authors are deeply grateful to the many people who have helped to bring this important story of KEEN Engineering to life – giving breath and warmth and light to these pages just as this leading firm does on every project in helping to transform North America's building industry and its practices.

First, we thank Kevin Hydes, KEEN's third and current president, for the opportunity to present this important body of work and for his vision in sharing the business methodology, design techniques and innovative "natural thinking" the firm has embraced to make the world a better place – one building at a time.

The *Ecological Engineer* could not have happened without the assistance, insights and careful guidance of KEEN Board of Directors Blair McCarry, Mark Mitchell, Sid Siddiqui, Met Ulker, Chris Jepson, and Paul Anseeuw. In addition, Jim Burns and his marketing team made it possible for us to feature so many wonderful photographs and illustrations of KEEN projects and concepts. Also invaluable as sources of information on topics that vary from KEEN's constantly evolving history to the Kyoto Protocol and the leading-edge of sustainable design were KEEN principals Jennifer Sanguinetti, and Jim Sawers.

This book is a testament, as well, to the company-wide KEEN spirit (sharp, vivid, enthusiastic) of those "KEEN-ers" who gave so generously of their time and expertise to convey the concepts and business strategies they use on diverse projects across Canada and the U.S. In particular, we benefited from the time and talents of KEEN principals and associates who included Bob Alexander, John Andary, Dr. Rosie Hyde, Mike Godawa, Tony Grice, Pasha Korber, Tom Marseille, Matthew Peachman, Dan Roberts, Mark Swain, Met Ulker, Tom Wilson and Matt Younger. Special thanks, also, to Sharon Lisette for locating important missing pieces; to Mary O'Brien for her tireless number-crunching; to Denise Tade for supplying an almost endless flow of information on projects; to Samantha Shah for her sharp edits; and to Claire Upton for keeping us always in touch with Blair, Jennifer and the rest of the much-traveled KEEN Concepts group.

Acknowledgements

KEEN would not be KEEN if it were not for the long and growing list of architects, engineers, planners, industry partners, supporters and others who have helped them to become North American leaders in sustainable design. Among those who provided us with valuable insights to the KEEN Engineering story were: Lynne Barker of the City of Seattle; Peter Busby of Busby Perkins+Will; Dr. David Cawood of David Cawood Consultancy Inc.; Bert Gregory of Mithun Architects; Eva Matsuzaki of Matsuzaki Architects; and Freda Pagani of the University of British Columbia.

Research, accuracy and access to information has been vital to the process of compiling the many facts and figures for this book. We appreciate the services and assistance made available to us through the Kansas City Public Library, the Vancouver Public Library, the Miller Nichols Library at the University of Missouri-Kansas City, and the public libraries of Edmonton, Regina and Calgary.

We are also grateful to our entire publishing team for *The Ecological Engineer*. Erin Gehle's inspired book design and her dynamic use of photographs, diagrams and illustrations have been essential to showcasing the true creativity and vitality of KEEN Engineering as a firm. Dr. Fred McLennan has been our keen editorial eyes while ensuring that the text remains, first and foremost, "a good read".

And finally, our biggest supporters have been our spouses. Thank you, Julie (Dave's wife), for your love and encouragement, your patience and understanding, and your shared excitement of the importance of this work and the beginning of this new journey in publishing. Thank you, Tracy (Jason's wife), for your continued support and encouragement, even while pregnant, allowing us to give birth to this important publication.

And to those we have forgotten, we promise to make it up to you!

—Dave Macaulay
—Jason McLennan

Foreword

by Bob Berkebile, FAIA

"Good engineers are hard to find," was my dad's response in 1970 to an experience I had shared concerning senior engineers I found to be frozen by the inertia of the way things "have always been." Since he shared that observation thirty-five years ago, the world of design and construction has undergone dramatic change – changes my dad would find unimaginable. But on this point he would still be right; good engineers are hard to find.

One of the questions I am most often asked is, "How do you find mechanical engineers who can do this?" The "this" they are referring to is high performance, integrated design. I always hesitate; after three decades of searching for engineers with these skills and experience, I can count the firms we utilize on the fingers of one hand. For me to share that the kind of engineer they seek are few and far between does not seem very helpful. So I usually point out that attitude is as important as experience. And for those who have difficulty with that concept, I always share the story about my first encounter with Kevin Hydes and KEEN Engineering.

Freda Pagani, from Campus Planning and Development at the University of British Columbia, decided in March 1993 that a series of new campus facilities that had just been funded should be designed to deliver a higher level of performance than they had been experiencing. She engaged us to work with the administration, faculty and local consultants on the design of these projects. The first was the C.K. Choi Building, an institute for Asian research. The architects were Matsuzaki Wright, led by Eva Matsuzaki, and the mechanical engineer was KEEN Engineering, led by Kevin Hydes. It was a very capable team even though only Cornelia Hahn Oberlander, the landscape architect, had any sustainable design experience.

I will never forget my first day with Freda and her C.K. Choi team. We spent the day in an educational forum on "the state of the campus and world in 1993" (as Freda and I understood it at the time) and how integrated, high performance design could improve the current conditions. As we finished the first day's work I asked if all participants would share, after dinner, the most important thing they had learned that day and what contribution they felt they could make toward achieving our client's goals. When we finished eating, Kevin was the first to speak. "I think I know what you want; you want a building without a mechanical system." I responded by confirming that a research facility without a mechanical system would be a worthy goal. Kevin's response was, "I don't have a clue how to do that but I could really get into trying to figure it out, if it were not for our contract." Their contract, like most mechanical consulting contracts at the time (and still very common today), was based on a percentage of the cost of the mechanical system. "So if we accept this goal, do a lot of extra research and are ultimately successful, our reward will be no fee." We agreed that our first task would be to find a more appropriate basis for KEEN's compensation.

The next morning, before we began the goal-setting session, I asked if anyone had experienced any new revelations or concerns about the information exchanged the previous day. Again, Kevin was the first to volunteer. After we finished the night before, he decided to visit the site or, more specifically, the steam tunnel beneath the street in front of the site before he left for home. He was excited to discover and share with us that the steam system was so leaky he was convinced we could capture enough waste steam to meet most of the needs of our building. And this was before he had a revised contract. We later discovered that the UBC Central Plant would not allow us to utilize this strategy, but it was a great way to begin a dialogue on goal setting.

I was impressed with the team's ability to participate in a collaborative dialogue of discovery and the team members' willingness to embrace high performance goals which no one, with the possible exception of Cornelia, had experienced before and which, for the most part, set new standards not just for UBC but for Canada and North America. The entire team deserves credit but especially Freda for clarity on behalf of the University, Eva for her leadership, and Kevin for his curiosity and enthusiasm.

In the first couple of charrettes I assumed that the openness, spirit of exploration and willingness to challenge anything other than comfort, health and efficiency that I appreciated in Kevin Hydes might be difficult to find in other engineers in the firm. But I discovered that this spirit was to be found in others at KEEN. In fact, a confluence of events including ownership changes and project experience was transforming the organization. The firm made a conscious decision upon Tom Johnston's retirement that McCarry, Mitchell, Hydes and Anseeuw would be equal partners and lead the firm, with Kevin as president. This collaborative leadership, coupled with new project experience at C.K. Choi, BC Hydro, Liberty Square and The Body Shop in Toronto, changed the firm's world view and launched the organization on a journey toward integrated design.

These changes empowered KEEN staffers like Jeanette Frost. One of the goals the team embraced for C.K. Choi was that there would be no sewer connection. The primary task of finding the best strategy for accomplishing this goal and selling it to UBC and the City of Vancouver fell to Jeanette. She was resourceful and tenacious in researching alternatives to conventional systems and in understanding the issues that motivated Dr. Crapper to create the water closet in the first place, which catapulted us toward today's conventional wisdom. She was successful in selling composting toilets for this multi-story building to all the authorities, and she seemed to thoroughly enjoy it.

Since then I have had the pleasure of working with Blair McCarry and others at KEEN on a number of projects. They are always consistent in their holistic, collaborative approach, and in their willingness to challenge the existing assumptions or rules in search of new levels of comfort and efficiency. I have enjoyed observing them utilize their core values and new capacity to deliver a healthy virus to project teams, often infecting other team members or the client to push existing limitations and consider new strategies and possibilities. During this time they have grown and transformed their practice while consistently influencing their clients to reach for more integrated, higher performance designs. It appears to be a very successful business model for them.

Currently, there are very few firms who approach design in this way but, fortunately, these few are leading an industry-wide transformation. And thankfully some were exploring and defining these opportunities before KEEN. The most encouraging thing about the movement, and the KEEN story provides ample evidence, is that a firm can change so dramatically from traditional practice to best management practice to an innovative practice that contributes to the redefinition of the field in just one decade.

KEEN is an impressive model of collaborative leadership, growth and financial success. But they are much more. Their investment of time and energy in sharing their discoveries and lessons learned with the industry through articles, lectures and workshops is extraordinary. Many participate in this endeavor but I have personally seen Kevin, Blair, and Jennifer Sanguinetti present and I know how effective they are as educators. Because my first contact was with Kevin, I know him the best. I have presented with him and I have observed his impact on audiences. And I have watched him provide critical leadership in the creation of the Cascadia Chapter of the U.S. Green Building Council, the first chapter of the USGBC, the Canadian Green Building Council, and now as Chairman of the USGBC.

Watching KEEN's dramatic transformation and influence over the last decade has given me hope. As a result, I am optimistic that this story of extraordinary transformation, practice and culture of KEEN will accelerate a healthy infection that will make this approach to practice more contagious.

My hope is that my dad's statement, "Good engineers are hard to find", will soon be obsolete.

Publisher's Introduction

KEEN Engineering

As someone who was trained as an architect, it has always been quite easy to find a host of beautiful books that provide inspiration and case studies on how to improve my craft. Architecture is visible and easy to explain through a combination of words, pictures and diagrams. Most architecture books are very graphically rich and accessible, even to the laymen who might not understand everything that they are seeing, but will, nevertheless, understand enough to either appreciate or dislike an architectural work.

The same cannot be said for the engineering profession. Most books written for engineers tend to be highly technical and something that must be slogged through in school as required readings or used as references for sizing ducts, fans, beams or columns. Few people outside of the engineering profession would choose to pick up an engineering publication for fun – not even some engineers. There are a few exceptions, but for the most part, the stock of engineering publications is not very appealing. In some ways this lack of appeal could be perceived as an outward manifestation of the public's perception of what it means to be an engineer- indeed, possibly the field's own self perception – that engineering is simply a high-tech field that is of little interest and that it should be relegated to the background of the built environment. After all, engineers typically only make the spotlight when something goes wrong with their work! We take note of the engineering craft when a building has a structural failure, suffers from sick building syndrome or is simply so hot or cold that we have to complain. When things are working well the engineers are all but forgotten, and their great work is all but invisible.

Over the past decade I have worked with dozens of engineers of various breeds and schools of thought in projects all over the country. I have been impressed by the intelligence and thoughtfulness of many, and I have learned to respect and appreciate the contributions that they make to all facets of the built environment. The critics are wrong. Engineering is sexy! It is a fascinating profession and deserves to be presented with as much fanfare and clarity as the architecture that cloaks it.

With regards to environmental performance, engineers carry a great deal of responsibility and at times play a larger role than the architect in lowering environmental impacts – from the way a mechanical system provides comfort, to how the building uses and handles water and wastes. Yet, despite this responsibility, the engineering profession has been slow to think beyond its primary responsibilities to serve their clients' functional needs to the wider responsibility of the environmental impacts these needs create. Innovations have been slow in coming for solutions that are both highly functional and highly responsible at all scales. Resources that teach this ethic to the various fields of the engineering profession are also few and far between. *Mechanical and Electrical Systems for Buildings* by Stein and Reynolds, and *Sun, Wind and Light* by G.Z. Brown begin this dialogue for MEP firms, but many more publications are needed to make their way into schools of engineering, to the shelves of professional firms and, perhaps most importantly, to cross over and engage the architects and the clients who hire them. Only when engineering is better understood, valued and simultaneously held to the high standard of environmental excellence can a future of sustainable buildings become possible.

With these issues as motives we, at Ecotone Publishing, are incredibly excited to be releasing this new series entitled *The Ecological Engineer* in which we hope to celebrate the most innovative engineers in North America and the practices and principles that they use to produce functionally outstanding structures, systems and technologies in a way that embraces the philosophy of sustainable design.

Through this series, we hope to help make the "invisible" visible and show the innovations and pioneering spirit to be found in the best practices and projects in our industry. Those engineering firms selected for coverage in our series make up a rare and inspiring collective that run successful practices by all conventional standards, but do so while embracing an ethic that is responsive to ecology – the ecology of people and place and natural systems.

These firms must meet four strict requirements:

- They accept and understand their broader responsibility as professionals to society and the natural environment.
- They are at the top of their practice by all conventional standards as well as "green standard".
- They actively seek opportunities to change and improve their profession in these areas.
- They cultivate a culture of innovation and resourcefulness with a strong vision of the future.

We are very proud to have selected KEEN Engineering as the first spotlight in our series. As mechanical, electrical and plumbing engineers who work on projects all over North America, they have consistently been inspiring on multiple levels – as a company, as individuals and by the sheer volume and quality of their work. As this book will show, KEEN is a new breed of engineering firm that embodies a critical mass of information related to responsible practice.

By focusing on what KEEN Engineering does – with practical examples – we hope to elevate the profession of engineering and contribute to a richer understanding of the relationship between the built and natural environments. Our book is organized into three important sections: The Practice – which will look at how KEEN as an organization is making change; The Principles – which looks at the most exciting processes, technologies, strategies and methodologies behind their work; and The Projects – which give in-depth analysis of five significant case studies, told this time from the engineers' perspective.

KEEN is the first example in the *Ecological Engineer* series and, as we explore their work and the work of others in forthcoming publications, it is a challenge to all of us in the building profession to get leaner, smarter and more sustainable with every project.

The future depends on us.

—Jason F. McLennan
CEO, Ecotone Publishing and Principal at BNIM Architects

Section One

The Practice

The Flyer, The Racer and The Pursuit of Excellence

A Brief History of KEEN

I am looking for a lot of men who have an infinite capacity to not know what can't be done.

– Henry Ford

Jim Keen loved to fly.

From Toronto to Regina, Calgary, Saskatoon and across western Canada – and later the United States, Australia, India, England and beyond – Keen was a born traveler, an innovator, an entrepreneur, a motivator. He would go where the work was. And his employees and friends often called him "Jumpin Jimmy Keen" for good reason.

In the late 1950s, Keen partnered with Jerry Yost to form Yost Keen and Associates in Toronto. The company successfully designed projects there, eventually making a name for itself as mechanical designers across western Canada. While working on the SaskTel Building, a new high-rise office for Saskatchewan Telephone in Regina, Yost and Keen decided they needed a bigger on-site presence to complete the project. So Keen, a native of Toronto who was born and raised in Cabbagetown on the city's east side, agreed to move west to Regina, joined shortly by Harry Hleck.

Courtesy of Keen Engineering

Jim Keen, "the Flyer" and founder, helped KEEN Engineering and its staff achieve early success through innovation and a non-stop work ethic.

Although the firm's growth continued steadily, in 1960 Yost and Keen agreed to split into two firms – with the Regina office becoming KEEN Engineering Co. Ltd. From their new vantage point in the prairie provinces, the fast-growing western Canadian market looked promising.

One of KEEN's first employees was Tom Johnston, a young Winnipeg native who had just graduated from Sheldon Williams Collegiate in Regina and was planning to enroll at a local college that fall to study engineering. After a series of temporary jobs, he was introduced to Jim Keen who had only recently arrived in town and Johnston was quickly hired as a student draftsman. He would stay with the firm for the next forty years. Another new KEEN employee and draftsman hired was Bob Alexander, who is still with the firm today.

It did not take long for KEEN to attract new business. In the early 1960s, KEEN designed the mechanical system for SaskPower's (Saskatchewan Power Corporation) new head office in Regina. In 1962, work began on the Centre of the Arts in Regina, a large, multi-year project that was completed in 1965 and opened in 1967.

Bob Alexander recalled the non-stop work ethic and camaraderie of those early days:

> It was just a good core group of people. We'd work for a month straight and then take a few days off. We used to get sandwiches and sleep in the building and go right back to the drafting board. Most people wouldn't have done that. But bottom line, we all respected each other. Everybody had a job to do and did it.

Flying High

Jim Keen founded his company on innovation and entrepreneurial prowess. He was a creator too, adept at starting up new offices to meet his clients' needs and always willing to fly off anywhere to find new business. In 1963, that willingness took him to Australia for the big Sydney Harbor project, where he quickly established an office in Sydney.

KEEN's next new office was closer to home, in Edmonton, Alberta. The office site was chosen in order to support a new major downtown office tower for Alberta Government Telephone (AGT). Overlooking the North Saskatchewan River, the AGT Building (now called Telus Plaza South) was the tallest building west of Toronto when it opened in July 1971. The AGT Building served as a stepping-stone to projects for the University of Alberta and the SaskTel (Saskatchewan Telecommunications) Building in Regina, an S-shaped high-rise structure completed in 1965. From there, KEEN designed buildings for the Western College of Veterinary Medicine in Saskatoon and the University of Saskatoon. By contrast, most of Keen's competitors were focusing on smaller commercial buildings.

"Jim Keen would go anywhere that anybody asked him to," said Alexander. Whether to Calgary, Seattle, Los Angeles, Sydney, or Boise, Idaho, he was prepared to hop on the next flight there. "Jim could never say 'No', that's the type of person he was. He liked to help people out, he loved to travel."

In early 1967, KEEN opened its Calgary, Alberta office to support work emerging there. Later that year, Jim Keen moved west to Vancouver to open an office when the firm was awarded the contract for 'Project 200', now known as Granville Square. This project was the start of an ambitious $1 billion downtown waterfront complex, overlooking the embarkation point for the Seabus linking downtown to North Vancouver. The office and retail building was located over top of the Canadian Pacific railway tracks and beside the old CP Railroad Station, erected in 1914.

By 1969, Calgary, Edmonton, Regina and other central Canadian cities were mired in a recession. Keen had discovered work overseas in New Delhi, India, for construction of the new Canadian Consulate building. At his request, Tom Johnston moved from Edmonton to the west coast to manage the Vancouver office. New KEEN offices appeared in Seattle, India and New Zealand. Despite the cost and limited resources, Keen also pursued an opportunity to open a U.K. office in London in 1972.

By the early 1970s, Vancouver had become KEEN's main base of operations, with a core group of engineers under Tom Johnston's careful, seasoned leadership. Healthcare architecture became a focus then, leading to a small but important project for the Lions Gate Hospital expansion in North Vancouver.

Mark Mitchell joined KEEN Engineering in 1973:

> I often think that much of the spirit of Jim Keen is still prevalent in the work we do today. I remember Jim saying to me when the oil embargo hit and everybody was trying to save energy, he'd say, 'Mark, it's not about saving energy, it's about saving money.' And I always tell people that, because it's a great story. People don't think about energy efficiency in maybe the right terms, they think about it in the most immediate terms.

Shortly thereafter, KEEN was hired to design the mechanical systems for thirty-five claims centers across the province for the Insurance Corporation of British Columbia (ICBC), a Crown corporation established in 1973 to provide convenient drive-up auto insurance, registration and licensing services to British Columbia motorists.

Due to market fluctuations in the late 1970s, KEEN closed its Regina office while opening its first in Victoria, British Columbia. Johnston continued to play a major role in the company as lead designer on a number of projects and special initiatives. He also continued to be a hands-on manager who stayed closely connected to the rest of the firm.

Along with Ken Junck of the Calgary office and Harry Hleck in Edmonton, Johnston approached Jim Keen about purchasing the firm. It was time for new blood and a new direction for the firm. They believed the time was right for KEEN Engineering to build on its thirty-year track record and take the next big step forward. Reluctantly, Keen agreed to the proposal suggested by the three partners.

New Direction, New Challenges, New Tools

Thomas E. Johnston loved to race fast cars.

His passion for motorsports dated back to those first days with the company when he raced small sprint cars at an old landing strip north of Regina. He continued to race for thirteen years, winning championships in 1963 and 1970. In all, Johnston's motor sports career spanned three decades as a driver, constructor and owner, competing in more than four hundred races at dozens of tracks across North America.

With the backing of the Tom Johnston Keen Engineering Formula Atlantic team, driver Frank Allers won the 1990 Player's Ltd. Canadian Formula Atlantic Championship and placed second in the series in 1989 and 1991. By then, KEEN Engineering had become widely known by clients, Vancouver residents and race fans across Canada for its sponsorship of high-performance race cars – and two scale models even adorned the walls of its Vancouver office.

Upon Jim Keen's retirement in 1980, Johnston was appointed KEEN's second president – about the same time as Canada's western provinces went into a heavy recession. With Johnston at the helm, it was clear he would be running things differently as a "stay-at-home" president. He was visible, approachable, often working closely with engineers on problem-solving. Mostly, he brought stability to the operations in the midst of a tough market as KEEN was forced to take on many projects that were marginally profitable. He also instituted new procedures for each office in order to better manage the affairs of the company. This approach began to move the company from its entrepreneurial roots to a more strategic focus on business.

Johnston initiated a series of regular, informal meetings with the entire staff to explain current happenings and what new opportunities were available for KEEN to pursue. These meetings became known simply as "The Fireside Chat" sessions.

"He had a way of being able to sit down with a bunch of people in a room, and he wouldn't be there as your boss, he'd be there as a co-worker," remembered Mark Mitchell. "He'd say, 'Here's what I think is coming down the pipe. Here's how I think we can get prepared for it. Let's talk about it.' Tom knew everyone and how they could respond to different challenges. So he was really good at being able to position the team."

KEEN's ongoing work for ICBC softened the impact of the recession somewhat. In 1981, mechanical work was completed on the SaskTel Building in Regina and Energy Square in Edmonton. Also in 1981, KEEN began work on the $22 million redevelopment of Vancouver General Hospital, the major patient care, teaching and research hospital in British Columbia – a relationship that would continue for more than 25 years. Other healthcare projects at the time included Jack Bell Research Centre, the Terry Fox Laboratory at the British Columbia Cancer Research Centre, and many other hospitals in the region.

> Sometime during our 35th year, we passed the $100 million in fees mark. That translates into $10 billion worth of building construction. I think that is pretty impressive. A really strong balance sheet, the best culture, the vision, the beliefs, the experience, the commitment, and the skills. KEEN is the best, let's keep it that way.
>
> – Tom Johnston, 1996
> KEEN's 2nd President

Courtesy of Keen Engineering

Tom Johnston, "the Racer", brought engineering discipline, professionalism and smart growth to his tenure as KEEN's second president from 1980 to 1999.

In 1982, Johnston and his co-workers seized another kind of opportunity that would have immediate and long-term benefits for the company. In those early days of computer-aided drafting (CAD), KEEN, like most engineering firms, still relied on paper, pencil and pen to draft technical drawings. That practice was changing rapidly with the introduction of the micro-computer in 1976, the first PC CAD software in 1979 and then the IBM-PC in 1981. Most CAD programs, however, were still extraordinarily expensive and usually operated only on turnkey systems. Johnston had been watching these developments closely, waiting for the right time, when affordable CAD hardware and CAD software would converge.

That moment presented itself at the 1982 AutoCAD trade show in Los Angeles. Most CAD manufacturers were still making the big machines. But Johnston found himself joining hundreds of other attendees in crowding around a single booth at the show. There, he saw a demonstration of how AutoCAD, the premier drafting package, could run on a small PC – for the price of about one thousand dollars. He quickly made the necessary software and hardware purchases, saw that every office had AutoCAD and knew how to operate it. Furthermore, he developed a plan for KEEN's staff to go out and teach local architects, their primary clients, how to use the software – a new incentive, an advantage of using the services of KEEN Engineering. "He was a real mastermind of the kind of philosophy required for marketing," said Mitchell. "Tom believed that was the only way out of the recession."

Again, A Boomtown

On May 2, 1986, the Prince and Princess of Wales invited the world to Expo 86, as the City of Vancouver was celebrating its 100th birthday. Among the sixty-five pavilions along the north and east shores of Vancouver's False Creek was the Ramses II Pavilion – containing treasures from the life of the pharaoh – and just completed by KEEN. By the time it finished in October, more than twenty-one million people had visited the world exposition.

Enthusiasm for the Expo was tempered by nagging unemployment and other gloomy economic news. Still, 1986 represented the first year in that decade when all KEEN offices made a profit. In October, Johnston asked Mitchell to open an office in Toronto, where they had had no real affiliations or activity since Jim Keen left there in 1960. By the end of the year, a new office was in place.

The move, recalled Mitchell, was intended to give KEEN a stronger presence nationally: "That was the focus, because people thought of us as a western company. So we wanted to give Toronto a shot, return back to the largest city in Canada and see if we could become a national engineering firm."

KEEN's Vancouver office was certainly eager for the bust to be over and a boom to begin – and it did, beginning in 1987. The next five years would be an impressive period of growth in the city's history, with an unprecedented surge in population and the buildings to accommodate them: houses, offices, laboratories, schools, courthouses, jails, big buildings, small buildings – many of them designed by KEEN.

By 1988, the Port of Vancouver was handling more imports and exports than any other port in North America. Both major universities in the city – the University of British Columbia and Simon Fraser – were expanding their campuses rapidly, and KEEN engineers formed close working relationships with the administrations, their architects and maintenance staffs. The firm was also selected to design mechanical systems for the new Fraser Valley Credit Union Building

Copyright © Robert Burley/Design Archive

The Body Shop-Canada

Location: Don Mills, Ontario

Type: Office

Size: Office is 67,000 sq. ft.;
Living Machine™ greenhouse is 4,000 sq. ft.

Completion: 1992

Architect: Colborne Architectural Group

Sustainable Features:

- Living Machine™ on-site system that treats the facility's domestic sanitary waste and production waste.
- Salvaged materials - several re-used plumbing fixtures from a nearby hotel renovation were installed in the building along with numerous fixtures from the previous head office and production facility of the Body Shop.
- Radiant heating - heat pump system with high efficiency gas-fired boilers
- Operable windows in office and work areas

and the new head office of the Association of Professional Engineers and Geoscientists of BC (APEGBC).

Making Conscious Choices

The early 1990s brought KEEN a remarkable series of opportunities – opportunities that signaled even bigger changes ahead – beginning with a new home office and bottling facility for The Body Shop Canada near Toronto.

For the innovative company founded by Anita Roddick in England in 1976, North America was the next logical place to grow its franchise network for its unique line of skin and hair care products. The Body Shop Canada launched its first store in Toronto in 1980, expanding to ninety-five locations coast to coast over the next decade. Eventually outgrowing its original headquarters north of Toronto, in 1990 the corporate partners needed more space but they also wanted to stay in the same industrial neighborhood where they had started.

The Body Shop Canada partners – Margot Franssen, Quig Tingley and Betty-Ann Franssen – declared that reuse and recycling would be a high priority for their next location. They chose to renovate and expand a 62,000-square-foot former publishing plant, just ten blocks away. In addition to many recycled materials, the building includes green features such as daylighting, energy-efficient lighting fixtures, and exterior insulation. KEEN was responsible for the mechanical system that includes a waste-heat recovery system. The system, in combination with other energy saving measures, have saved the company an estimated $45,000 in energy costs annually, resulting in a two to three year payback on the capital cost of the insulation, plus a $250,000 savings on the purchase of a smaller heating and cooling system.

> We were essentially living an experiment with underfloor air. In our new office, we discovered that if the air conditioning failed, the building didn't get particularly warm in the summer. By blowing air into these floor spaces, we're really using the power of the building to cool the building – so we were starting to understand thermal mass. We were beginning to explore a lot of things that we didn't understand yet.
>
> – Kevin Hydes, KEEN

Another significant feature of the new headquarters is its Living Machine™ – the first installed anywhere – designed and constructed by biologist Dr. John Todd. Contained in a greenhouse on the south side of the building, this closed-cycle treatment system reclaims 4,000 gpd of wastewater from the main distribution and bottling facility, as well as sewage created by the company's one hundred employees.

Equally significant for The Body Shop Canada and KEEN was the creation of a "green decision-making matrix" created expressly for this project. In planning its new headquarters, the partners sought ways to express The Body Shop's core values of social responsibility and use of environmentally-sustainable resources. Yet they also felt overwhelmed by the hundreds of different options presented to them by the consultants and environmental services firms they had hired.

"The owners were trying to do the right thing," recalled Mark Mitchell, "but they couldn't always determine which design or system choices they really preferred for the building. So we came up with the idea of this matrix, a way to make their choices compelling, one solution in favor of another, to best fit their particular type of business."

Not only did the new decision-making tool expedite the process and help define strategies that best fit their business needs, it put KEEN in direct contact with the owners. The project also gave KEEN a very early introduction to the fundamentals of green buildings. And the green decision-making matrix would be used, refined and incorporated into many later projects.

The Air From Below

In Vancouver, while much of the building industry in North America still faced a recession, the local economy was on fire and KEEN worked feverishly to keep up with the demand. A major commission came with the design of mechanical systems for the new 500,000-square-foot corporate offices of BC Hydro in Burnaby, British Columbia. As BC's major electric utility, the company wanted a state-of-the-art, energy efficient example to the province for its new eighteen-story office tower, which also needed to accommodate a high rate of churn as staff were moved frequently. KEEN recommended an underfloor air distribution (UFAD) system for the BC Hydro Edmonds Building, which would represent the first and largest commercial use of an underfloor air system in North America.

KEEN's lead designer on the project, Blair McCarry, had some limited experience with access floors, including a large underfloor wiring project for the Workers' Compensation Board Office Building in Richmond, near Vancouver, in the early 1980s. A proposed new office for MacMillan Bloedel, the lumber giant based in Vancouver, presented another early opportunity in 1985 to pursue underfloor air when design specifications called for an interior atrium free of ducts and other mechanical components. McCarry prepared initial concepts and an underfloor air schematic, but the project was canceled.

The BC Hydro project would involve fast-track construction totaling more than $300 million. Design was underway, even as KEEN's Vancouver office was in the process of moving into its new building in North Vancouver. When BC Hydro's project manager, Norm Webster, learned about the move, he asked: "If you guys think underfloor air is such a good idea, you should do it yourself." So, just as the foundation was being poured for their office, KEEN's senior management stopped the contractor and immediately weighed the feasibility of underfloor air for the new space. A quick decision was made. They designed the scheme, paying the premium for stopping construction and, subsequently, they had an access floor installed. Soon they moved in.

BC Hydro Edmonds Building

Location: BC Edmonds Center, Burnaby, British Columbia

Type: 18-story Office Tower

Size: 290,000 sq. ft. and two 3-story "podium" buildings of 100,000 sq. ft. each

Completion: 1992

Architect: Stantec Architecture

Sustainable Features:
- One of the largest access floor systems installed in North America
- Lighting and daylighting sensor controls
- Ultra high efficiency motors to all HVAC equipment
- Free cooling chilled water heat exchanger

"It became the ultimate test," said Kevin Hydes, who had joined the firm in 1982 and was manager of KEEN's Vancouver office. "We realized that Norm was definitely right, that we had to step up in this situation, asking ourselves: if this is the great system of the future, why are we not doing it? So we decided we'd bite the bullet and make the investment. It wasn't until we had installed the underfloor system and moved in – while the entire design team came and literally 'kicked the tires' – that this decision became the tipping point on the BC Hydro project; the fact that we were able to experiment on ourselves. The overall consensus was that this was a sound, progressive approach and the benefits were clearly there for everybody to see."

Another big breakthrough on the BC Hydro project came when an internal analysis by the utility determined the underfloor air/access floor approach would require a significant incremental cost. But the analysts also calculated the average yearly churn rate of 50 percent in moving staff and the cost of ongoing operating expenses such as to move partitions, and bring in electricians. When factoring in how quickly workstations could be reconfigured, access flooring became an easy decision for BC Hydro as owner-occupier, because it represented just a two-year payback period.

Soon after completion, the BC Hydro Edmonds Building as well as KEEN's new Vancouver office would be visited regularly by architects and engineers from across North America – and frequent tours continued for the next two years.

"Inspiration Through Information"

On August 14, 1992, the Vancouver City Council announced the winning submission in its international design competition for a new, state-of-the-art central library. Library Square, as conceived by Moshe Safdie & Associates and Downs/Archambault & Partners, bore a strong resemblance to Rome's famed Colosseum – and has since become one of downtown Vancouver's defining landmarks.

Copyright © Jim Burns

Library Square

Location: Vancouver, BC

Type: Library

Size: 390,000 sq. ft. library; 390,000 sq. ft. office building; 30,000 sq. ft retail space

Completion: 1995

Architect: Moshe Safdie & Associates/Downs Archambault

Sustainable Features:

- Underfloor supply air using stratification
- High efficiency motors
- Library concourse heated/cooled by the library relief air.
- Dynamic thermal storage (thermal mass)
- Low temperature supply air and ice storage

(See Section Three for a detailed description of this project.)

The $100 million multi-building complex covers an entire city block and consists of the nine-story library, a twenty-one story tower for provincial and federal government offices, and a soaring, five-story glass atrium. The atrium, an enclosed, unheated concourse spanned by bridges, highlights the space between the library wall and the curving outer wall and provides a walkway between two city streets. Equally innovative at Library Square are the mechanical systems, integrated by KEEN into the design of the buildings and complying with the requirements of ASHRAE Standard 90.1. A key element of the library's interior space is its underfloor supply air system. The system uses a low-pressure, floor plenum approach with an access floor consisting of 2-foot by 2-foot removable floor tiles – which, in turn, provide easy access to computer cables and electrical wiring inside the floor plenum. The underfloor air system also integrates well with the rigid frame concrete seismic bracing required for the structure.

Once again, McCarry served as KEEN's lead designer: "The raised-floor space acts as a supply-air plenum, delivering good ventilation where people are located. We think this approach also significantly reduced the space cooling load, allowing heat from people, lights and computers to stratify – something you don't see in conventional buildings."

KEEN provided several other innovative design features for construction of Library Square. Much of the exposed concrete structure, expressed as architectural features such as the ceiling/floor slabs, absorbs interior radiant heat and further reduces the peak cooling load. An ice storage system supplies low-temperature supply air for the library. As a result, this primary air ensures a drier interior environment for the preservation of books and lower operating costs. The entry concourse has significant glazing and skylights and it is not heated or cooled directly; instead, the space is ventilated by relief air from the library. In the tower, operable windows provide high and low ventilation during moderate weather.

KEEN's work on Library Square was recognized with the prestigious ASHRAE Technology Award in 1998, gracing the cover of that year's March *ASHRAE Journal*. Equally significant, though, was the impact of the project on KEEN's design evolution as a firm. McCarry and others were now exploring the influence of a heavy structure on thermal mass. They were also starting to better understand stratification and experiment with techniques for moving cooler air from floor to ceiling to manage comfort. "We were starting to really look at the optimization of our engineered systems," said McCarry. "Instead of a big chiller for cooling, we used a small chiller and ice storage system to produce low temperature air overnight which was then released during the day. It was a pretty big move. Combined with the underfloor air, we were able to reduce electrical demand charges and downsize the air handling system and ducts to achieve huge cost savings for the project."

After Library Square – with the knowledge and momentum gained as a result of projects like the C.K. Choi Building, BC Hydro and The Body Shop – KEEN's phones kept ringing.

Another Glimpse of Green

While completing the BC Hydro project, moving into its new North Vancouver office, and finishing a series of secondary schools around Vancouver, KEEN was operating in high gear. The firm had received a Canadian Consulting Engineers Award of Merit for its recent work on Saanich Commonwealth Place in Victoria, BC. And Library Square was already underway. Then came "The Choi". By late 1992 and early 1993, the Vancouver market was still booming. Speed of delivery was imperative in applying mechanical solutions to each new building. Often that essential meant quick, instinctive decisions about systems: this building needs a VAV system, that one is a fan coil job. Vancouver's builders typically wanted cost-effective "commodity" designs that they understood. Replication was essential too: mechanical designs tended to look a lot like the designs used in the last building.

The market also demanded "technology" – and KEEN provided it. CAV-VT systems, supply air temperature sensors linked to energy management controls, pipe fan coil units used to draw return air through specially-built duct chases. Standard industry doctrine called for applying more, not less, technology to buildings, regardless of orientation, the building envelope or the architecture. If it was too hot or too cold, most engineers would add a micro-switch next to a window. Too much air or too little? Change the AHU configuration. Add more computerization, deliver more information.

Kevin Hydes recalled his own practices:

> I think I was probably designing the most complicated, most energy-intensive buildings on the planet. I mean, these were good buildings – but complicated – because we'd added too many devices, too many systems, too many subsystems. People were asking for speed, not efficiency. And more often than not, we were faced with a difficult period in commissioning certain buildings. Because we had less time, we often had to add more things, so it started to become more and more difficult.

Opened in 1996, the C.K. Choi Building at the University of British Columbia, Vancouver, became one of North America's greenest buildings – and later launched KEEN down its new design path.

KEEN's relationship with the University of British Columbia (UBC) dated back more than five years to the design of several buildings on its campus. By 1993 the institution, located in southwestern Vancouver near the coast, was still undergoing an aggressive phase of building and expansion. Yet, the new C.K. Choi Building for the Institute of Asian Research would be like nothing else on campus, and certainly like nothing KEEN's engineers had worked on before. The University's letter of invitation to the firm stated simply that the building "would be a demonstration project on sustainability for UBC."

On a rainy day in March 1993, Kevin Hydes' engineering career – and even KEEN's direction as a firm – was about to change. That is when he arrived at UBC to attend the initial design charrette for the proposed C.K. Choi Building and met Bob Berkebile, a smiling, 53-year-old architect from Kansas City, for the first time. Berkebile, as founding chair of the AIA's Committee on the Environment, had been asked to lead a two-day session so the project team could share joint goals and set objectives. Hydes knew many other participants in the room, including the campus engineers and Freda Pagani of UBC's Campus Planning and Development Department. He was introduced to the project architect, Eva Matsuzaki of Matsuzaki Wright Architects. This unique, multi-disciplinary charrette required that all key stakeholders attend as well: the structural and electrical engineers, the landscape architect, and any other contractors and consultants who would be contributing to the project.

By the end of the charrette's first day, after a number of sustainable design features had been identified – including daylighting, natural ventilation and composting toilets – Hydes addressed Berkebile directly. "I think I know what you want," he said. "You want me to design a building without a mechanical system, that's what I think you're trying to say." "I think that sounds about right," responded Berkebile.

Hydes later described the stated design goals for the Choi as "lifting a big three-ton load off my shoulders." For him, it represented a return to basics, a challenge to KEEN to design solutions that focused on energy efficiency and to think very differently about occupant comfort.

Later, he reported back to Tom Johnston: "Tom, I think we just got the job that'll change the future direction of KEEN." "But he was pretty nonplussed about it," recalled Hydes, "comparing it to the '70s oil crisis and the effect it had on energy efficiency in buildings. I remember him saying this new idea of sustainability 'will come and go' too."

Yet the pure energy of that charrette and the unflinching commitment of the University and project team drew Hydes and a few others at KEEN completely into the project. Instead of following the usual systems-intensive approach with big chillers and complex controls, he began to draft a series of schemes showing how the natural ventilation, occupant controls and other details could be achieved. In a matter of weeks, the outline of the building started to take shape: a narrow, thirty thousand square-foot structure dominated by a series of swooping metal roofs and a brick exterior featuring a number of operable windows and fresh air vents. Five atria below the curved roofs created a stack effect for natural ventilation. The building would also rely on natural ventilation for cooling, while the atria, windows and narrow floor plan ensured large amounts of daylighting.

The challenge for KEEN was to design and build a system that was not mechanical. But it also challenged their own HVAC beliefs – and those of UBC's engineers – that every building has to be mechanically ventilated.

"Right at the outset, our diagram showed natural ventilation would be an ideal, low-intensive approach for the building and would also meet its sustainability goals," said Hydes. "At first, the campus engineers said 'no' to naturally ventilating the building, until we forwarded a copy of the code with the word 'naturally' circled in red." And because KEEN engineers could not validate its approach other than with historical references, they relied on empirical data, old design manuals and European references – anything that might provide some insight to design questions about elements such as temperature and ventilation rates to help demonstrate code compliance. Manual calculations were confirming KEEN's design: how many people would be in the building, how many cubic feet do they have to breathe, how long would it take for the building to even hit this threshold.

"We were starting to explore what the codes were based on in the first place," said Hydes. "So, effectively, we were re-learning. We were having to go back and re-educate, re-confirm, re-understand ourselves."

Upon the official opening of the C.K. Choi Building in October 1996, the new center was introduced as a "model of possibilities" toward a sustainable built environment and the "embodiment of new standards for sustainable design." It was also hailed as one of Canada's most environmentally-sound building projects and it would soon gain numerous international awards for the common vision achieved by the University and project team.

Low-Tech, High-Tech in Parallel

Just as Hydes was realizing the tremendous potential of a low-tech path on future projects after the C.K. Choi Building, Blair McCarry continued to refine his high-tech ideas on KEEN's large building projects – exploring optimization and ways to further improve efficiency. Their ongoing debate and explorations were taking Hydes and McCarry down two distinctly different, yet parallel, paths in achieving clients' goals. Yet Hydes' intuitive designs and McCarry's hard science approach often met in the middle, each process driving them toward the same general solution.

In late 1994, KEEN partnered with Arthur Erickson of Architectura to design the Downtown Campus for BCIT (British Columbia Institute of Technology). Located in a busy urban neighborhood in Vancouver, the 168,000-square-foot educational project was notable for several reasons. KEEN's own knowledge base about ventilation rates and ventilation effectiveness had grown exponentially in just a few years, allowing the firm to put a variety of system choices before clients that were more focused on health and air quality issues.

Once constructed, the $40 million BCIT campus featured mostly traditional systems but it also had a high-tech ventilation scheme to achieve 100 percent outdoor air free cooling – resulting in nearly twice as much ventilation as any other building anywhere in Vancouver at the time.

KEEN went on to design several more community colleges in Greater Vancouver, including Douglas College – all large building projects, all featuring high-tech, high-efficiency solutions. By then, KEEN designers were starting to consciously chisel away, often without being asked, at finding more efficient solutions for every project. At the same time, KEEN's high-tech design proponents were increasingly sharing their ideas about underfloor air in offices with the firm's low-tech designers who tended to specialize in university and educational projects.

In 1996, KEEN's Edmonton office formed an important alliance with Rose Technology Group (later to become Vestar Energy Group), one of North America's largest performance contractors. Over the next several years, this alliance would allow KEEN to expand its services to include energy project identification, commissioning, monitoring and verification, energy audits, energy modeling and more.

That same year, McCarry traveled overseas to attend the annual CIBSE (Chartered Institution of Building Services Engineers) conference in the U.K. There, he learned about the latest techniques in natural ventilation and emerging standards in the U.K. Tours followed to a number of famous, recently completed, natural ventilation buildings in England. These included the Inland Revenue Headquarters in Nottingham, the Queens Building, School of Engineering and Manufacture at DeMonfort University in Leceister, the Anglia Polytechnic University Learning Resource Centre in Chelmsford, the Ionica Telecommunications Building in Cambridge, and many others. In building after building, McCarry was able to observe the techniques and strategies used in these leading examples, many included in the CIBSE applications manual: natural and mixed-mode ventilation,

Revenue Canada

Location: Surrey, BC

Type: Office

Size: 144,000 sq. ft

Completion: 1998

Architect: Busby Perkins + Will

Sustainable Features:
- Exceeds ASHRAE 90.1 standards by 30 percent
- Natural ventilation – operable windows around the perimeter
- Motion sensors linked to heating to control internal temperatures
- Underfloor air system
- Thermal mass
- Ozone-friendly refrigerant
- High efficiency boilers
- Storm system design with a retention pond to decrease the impact on city storm drains

daylighting strategies, high levels of thermal mass, and low-energy design approaches.

In late 1996, at the invitation of a local Vancouver supplier, McCarry was off again, this time to the Krantz Komponenten plant in Aachen in Germany's Rhine Valley. Krantz remains a leading manufacturer worldwide of innovative cooling and heating systems and products. For McCarry, the trip essentially became a week-long training course in European design techniques – specifically, underfloor air, air displacement, chilled beams and chilled ceilings. He also toured a half dozen installations featuring Krantz products at work in banks and other buildings across northwest Germany.

"I really picked up on a lot of the European design criteria, such as the very different comfort temperature standards and rules in place in Germany," recalled McCarry. "It opened my eyes, solidified my thinking to a whole different way of approaching building design."

Where High Efficiency and Reliability Meet

KEEN's achievements and growing reputation for designs on Library Square, C.K. Choi and BC Hydro opened the door to a whole new series of innovative projects. In partnership with architect Peter Busby of Busby & Associates, KEEN submitted ideas for the design-build competition on the new Revenue Canada office in Surrey, British Columbia. At $15 million, economics were a big factor – and the BC government had very clear energy targets in mind. Three teams of architects, engineers and contractors presented their ideas to the agency.

Busby and KEEN won the competition. Their concept of the "advanced office building" featured an underfloor air system plus a number of pre-LEED strategies: external shading devices, daylighting, indirect lighting and low density lighting. The building operates at well below (60 percent to 70 percent) targeted ASHRAE 90.1 compliance. It became a compelling example of green design principles while staying within a competitive office building capital budget.

"By now, we were really squeezing down the size of the equipment – not using any industry manuals we could find, only the design knowledge we had gained from these other buildings," said Hydes. "Again, this was adding to KEEN's own research and development efforts. We basically had a system that I'm sure was half of the size anybody else would have been installing."

KEEN's advanced building design techniques for the new, five-story tax center aimed at providing employee comfort, work space flexibility and energy sustainability. The large floor plate is 95 percent efficient and 90 percent of workstations are within twenty-five feet of glazing and natural light. Operable windows and the absence of conventional dropped ceilings give all employees access to natural light and ventilation in high ceiling spaces. Personal control of air supply is available at each workstation. Sunscreens, light shelves and clear low "E" glazing allow deep penetration of natural light, filter solar gain and further reduce operating costs.

"It became our challenge, really, to demonstrate that our calculation methods were in fact actually better than traditional methods," Hydes added. "But it was a huge success, because we proved – to ourselves and to the industry – that we were oversizing these systems dramatically. We had now finally designed a building with an underfloor system at the same price as a regular building."

Following the Revenue Canada project, KEEN projects started to appear in a growing number of industry publications and even in the pages of technical literature distributed by access floor manufacturers.

Heading into 1998, the stage was set for even more innovative designs emphasizing improved energy efficiencies, better daylighting and more natural ventilation. KEEN buildings were starting to function much more in sync with the natural cycles of day and night, which meant smaller equipment, less ductwork and more occupant controls – and comfort.

Green Project Tree

The branches of KEEN's "Green Project Tree", influenced by low-tech or high-tech design thinking, have become increasingly intertwined – as its engineers leverage internal expertise to include both types of systems in more buildings.

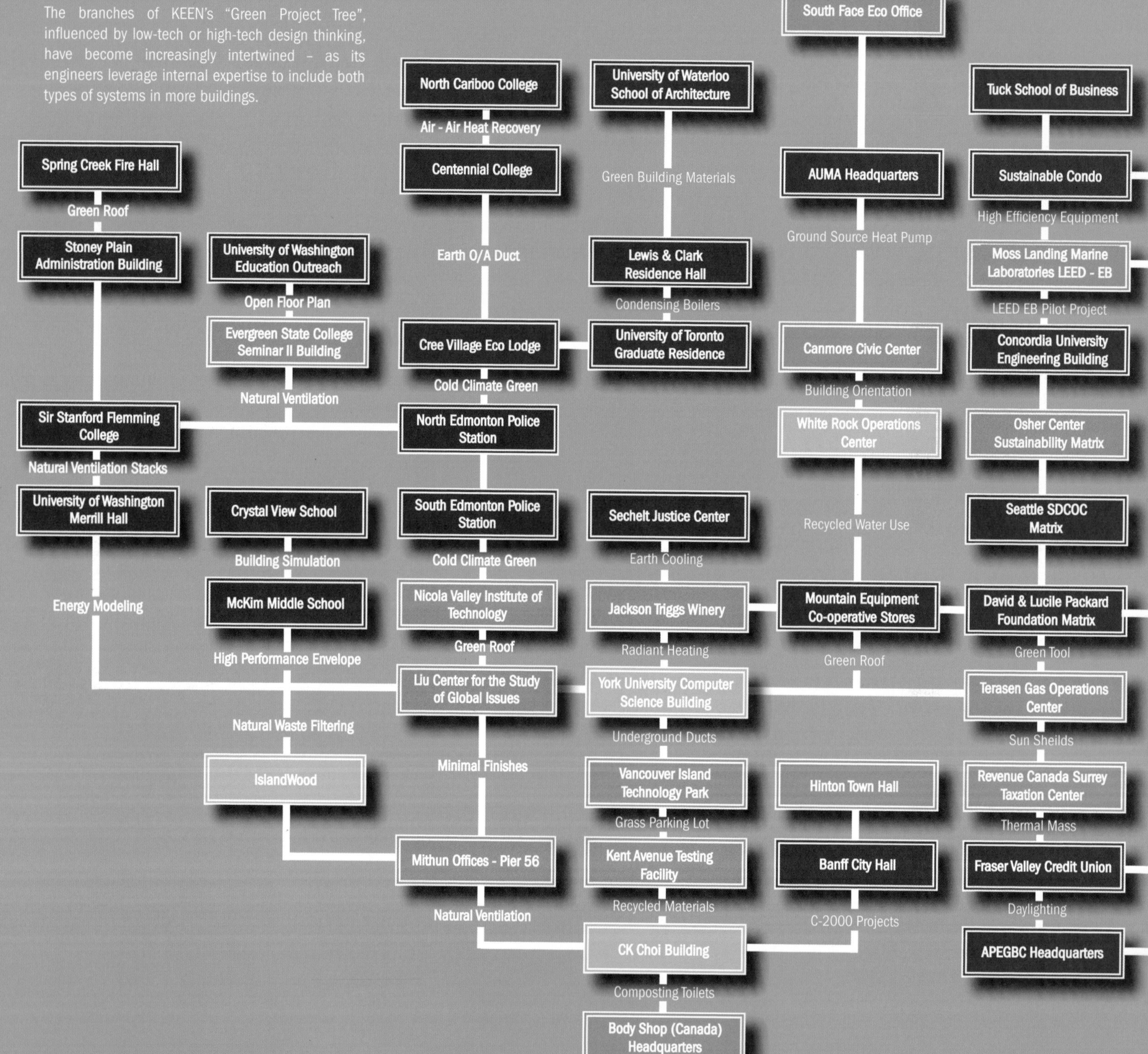

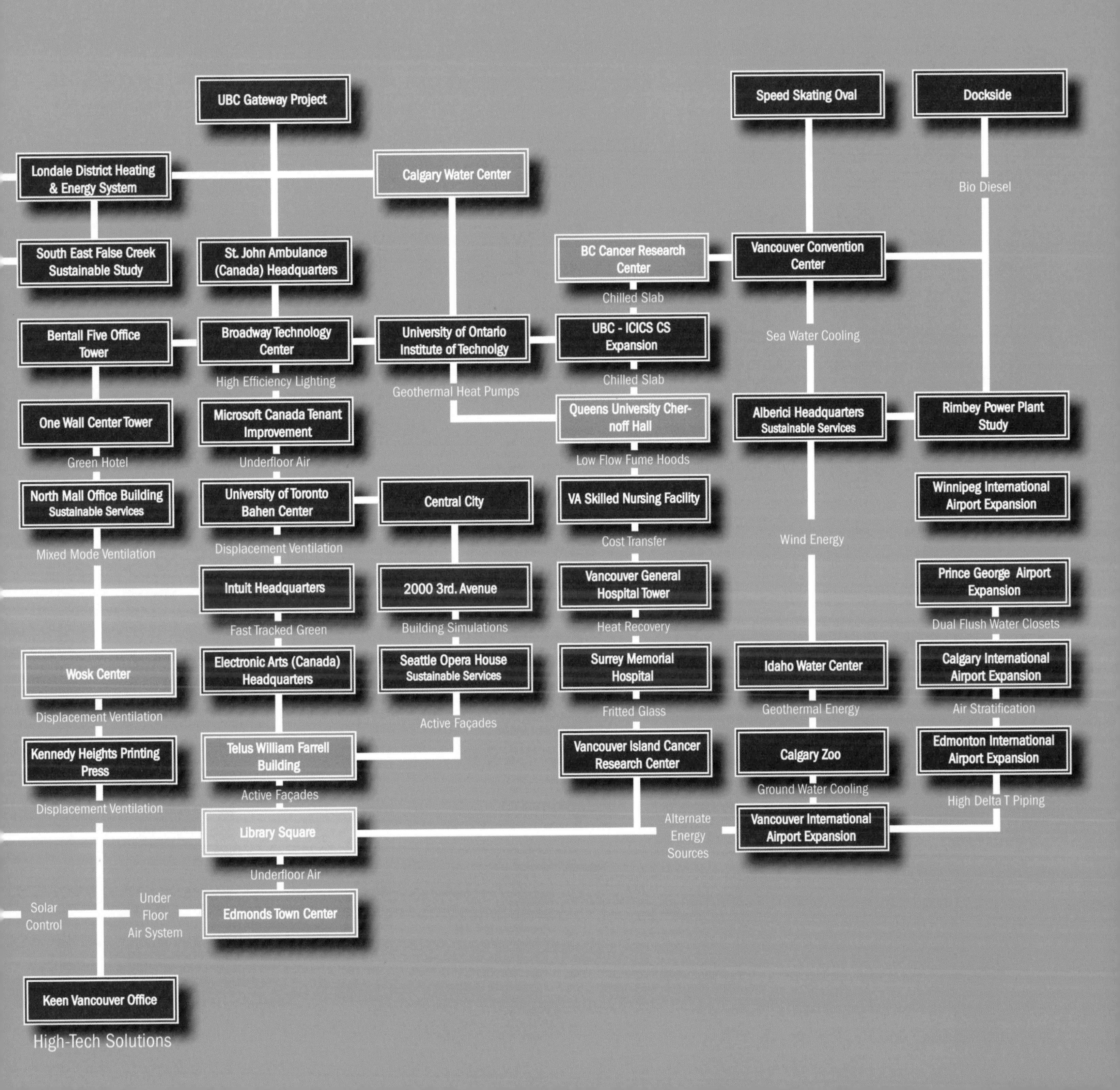

High-Tech Solutions

From Blue to Green

Transforming KEEN's Business

Man is small, and, therefore, small is beautiful.

– E.F. Schumacher, *Small Is Beautiful: Economics as if People Mattered*

By the late 1990s, KEEN Engineering was continuing to build its reputation for underfloor air systems and explore other alternative heating and cooling techniques. Internally and externally, the firm's design vernacular, too, focused increasingly on terms like "minimization", "optimization", "simplicity" and even "elegance".

The company's ongoing transformation from KEEN "Blue" to KEEN "Green" was not planned – and was often subtle, appearing and disappearing in unexpected ways. The transformation was also organic and decentralized, with distinct, disconnected activities happening independently in Edmonton, Toronto, Vancouver and elsewhere.

KEEN had always taken pride in approaching its projects in a collaborative way. But some of the firm's designers now realized that these alternative air, heating and cooling systems were causing them to re-think their ideas about collaboration. They had excelled at satisfying the needs of architects. Also, they understood the demands of owners or their representatives in ensuring that projects were delivered on time and on budget. But they had too often overlooked the third leg of the tripod: the operators with responsibility for actually managing and maintaining the buildings.

"We were beginning to get a better understanding that greener designs required us to address the specific needs of different groups," said Mark Mitchell, "whether that involved financing the building, energy strategies, or the cost to operate it effectively. To be successful as an engineer and as a practice, we were recognizing that it took all three components – the architect, the owner and the operator – to contribute equally to a successful outcome."

Kevin Hydes was seeing the same thing:

> We thought we had a good system and effective approaches, and then we'd hear from the maintenance guys about a problem on the last job. So we would add another system to fix it, and

Copyright © Jim Burns

Morris J. Wosk Centre for Dialogue

Location: Vancouver, BC

Type: Conference Center

Size: 42,000 sq.ft.

Completion: September 2000

Awards: 2001 Province of BC Heritage Award of Honour
2001 City of Vancouver Heritage Award of Honour
2001 UDI Award of Excellence of Heritage Redevelopment and Mixed Use Redevelopment

Architect: Stantec Architecture

Sustainable Features:

- Demonstrates the creative use of a heritage structure
- A displacement ventilation system which supplies air at floor level
- Air grilles are equally distributed to give optimal air quality at occupied zones
- The atrium serves as the return and relief air plenum for the entire building, reducing the amount of air required to temper the atrium space, thus reducing the energy load and operating costs

sure enough, if there was another problem, we'd add something else again. It's interesting, because the operators were not asking for a smaller piece of equipment that would be easier to maintain. In fact they said, 'Give me the big stuff.' So we were just adding more complexity, throwing more technology at the problem, not dealing with the root of it. As KEEN's work continued to grow exponentially, I think that a gap, a delamination, was taking place on the operational side. These were the people who ended up operating these overly-complex systems and potentially wasteful energy solutions.

Deeper Shades of Green

In the second half of 1997, KEEN had begun work on several projects in Vancouver and Toronto, each requiring of the firm's designers new and different ways to approach a mechanical solution – with equal parts invention, consultation, intuition and analytical skill.

Vancouver's new Morris J. Wosk Centre for Dialogue would be dedicated to understanding effective communication and serve as a physical expression for important international policy debate. Located in the heart of downtown, renovation on the forty-two thousand square-foot, five-story, former 1920 heritage bank building began in July 1997. KEEN joined Architectura Planning Architecture Interiors Inc. in creating the world-class conference and meeting facility as part of the Harbour Centre campus of Simon Fraser University.

The extensive re-design ultimately resulted in a wide range of meeting rooms and the unique Dialogue Hall. With its circular seating format for 150 people, the Dialogue Hall was designed specifically for face-to-face discourse. KEEN designed a displacement ventilation system for the Hall to supply air at the floor level for conference attendees. Another design feature is the four-story entry atrium – fully interconnected with the Delta Hotel next door – that also serves as the return and relief air plenum for the entire building, thus reducing the energy load and operating costs.

Since its opening in September 2000, the center has been the catalyst for groundbreaking discussions and seminars on issues ranging from restorative justice to real estate, health care to transit, and international law to art. A few months after the center's opening, KEEN's Toronto office was contacted about designing mechanical systems for the new $40 million Chernoff Hall Chemistry Facility at Queen's University in Kingston, Ontario. Brisbin Brook Beynon Architects of Toronto specialized in world-caliber chemistry buildings and were familiar with KEEN's work on a chemistry laboratory at the University of British Columbia during the mid-1980s. The design there was consistent with the architect's own approach, and KEEN won the contract.

KEEN's main challenge for the five-story, 140,000 square-foot building was ensuring the safety of occupants in the research labs. Designers explored several methods to maintain sufficient make-up air supply for fume hoods without sacrificing occupant comfort. "The University was completely clear about their goals for the project – when you're talking about chemistry and lab hoods, safety is obviously number one," said Mitchell. "So when we started talking about risk, it seemed to elevate our position with them. After all, it's simply a matter of putting the client's hat on, being able to see it from their point of view in order to get the business case right. At the same time," he continued, "we had previously worked on a number of these laboratory buildings, and we knew they could be very energy intensive. There can be a big energy cost, especially in cold climates, because a large volume of air needs to move in and out of the facility."

KEEN then discovered an environmentally-friendly solution for Chernoff Hall: a low-flow fume hood produced by a Long Island manufacturer. Together with the architects and key stakeholders, including the dean of chemistry and the university director of facilities, KEEN's designers visited the manufacturer to see the concept first-hand. They went on to the inventor's lab in New Jersey and then to several installations around Philadelphia where the low-flow fume hood was in use. They were convinced it was soundly engineered. With a smaller dimension than conventional laboratory hoods, the low-flow design would dramatically reduce the amount of air ducted out of the building – resulting in half the energy, half the expense, and much less maintenance due to smaller systems and fewer moving parts. "If we hadn't gone to that effort of researching the product, we would have spent all of our time on the education process," said Mitchell. "Otherwise, there was no way we would have installed and configured those particular fume hoods. We all knew it was absolutely the right thing to do. That's another example of the green decision-making process – whether you're considering a low-flow fume hood or any other strategy different from a standard approach. The client made the decision with us and chose to take a risk."

As the building design progressed, KEEN conducted an indoor air quality analysis to determine the best air distribution strategy for the labs to ensure user safety. As a result, fume hoods were located away from exits and supply air diffusers. The laboratory air system provided 100 percent outside air for maximum air quality. High-plume staged exhaust fans were used to filter and dilute the contaminants from the fume hoods before exhausting to the atmosphere. Several energy efficiency strategies were also applied throughout the building, designed to exceed ASHRAE 90.1 standards: underfloor air supply in the lecture theater; district steam incorporated into the perimeter hydronic heating design via radiant panels; exposed ceilings in the laboratory wing to create a thermal flywheel effect and bounce daylight into interior spaces; thermal massing with exposed concrete in the lab wing to absorb and release heat to offset peak loads; and stormwater retention techniques, including catch basin storage for irrigation and flow-control roof drains. Upon its completion in 2002, the Queens University Chemistry Lab became the largest installation of low-flow fume hoods in North America. The next year, the building was recognized by *R&D Magazine* as the "2003 Laboratory of the Year".

Before year-end, KEEN also started work on the new BC Gas Operations Centre in Surrey, British Columbia. Because of its location in an earthquake zone, the 180,000-square-foot office facility for

BC's major gas utility would be designed as a post-disaster building. BC Gas also wanted the high-tech office space and other buildings on the campus to demonstrate the potential of an energy-savings design to its gas customers.

KEEN's Blair McCarry teamed up with architect Mark Whitehead of Musson Cattell Mackey Partnership (Vancouver) to design the new building. Surprisingly, the utility did not provide any specific energy goals or targets for the open-plan office space. So the project team started with the basic elements of a low-energy building, and McCarry incorporated some of the design concepts he had studied in Europe, particularly the naturally-ventilated Ionica headquarters in the U.K. "It was a good process to go through," said McCarry. "We got the fundamentals down quickly and began working out the details, considering the possibilities of underfloor air, thermal mass, maybe some natural ventilation or a mixed-mode scheme."

The Operations Centre was divided into four segments, surrounding a central atrium for airflow and daylighting; a large skylight was placed in the middle of each of the modules. Daylight penetration was maximized by increasing the floor to floor height and providing extensive glazing. In addition, the designers oriented the building on an east-west axis to provide maximum exposure on the north and south façades. The east and west shear walls are massive architectural walls with punched windows. Solar control for the windows in these walls is provided by means of "Sun-Shields," large sheets of glass attached externally to the wall that allow the brightness without the direct sunlight. The south wall includes sunshades on the external face as well as light shelves on the inside. The north exterior is clear glass. Indirect lighting was used throughout the office floor areas.

The team's final design also optimized the use of natural ventilation to include operable windows, 50 percent of them operated by the building's DDC system. An underfloor supply air system provides displacement ventilation. Openings in the atrium walls allow fresh air to be drawn into the building and migrate by stack effect to the top of the atrium where the air is exhausted outdoors. The building was also designed for "night purging," allowing the ventilation to cool the structure at night and thereby reducing cooling loads during the day. The campus also includes a large landscaping pond, which is utilized as a water reserve. The pond serves as a filtration system for the stormwater system and bio-swales and as a fire protection back-up in the event of post-disaster water main failure.

McCarry was pleased with the process, the collaboration and the end result:

> It was a good interplay back and forth with the architect, working through the options, pushing and testing ourselves. I think the interaction on that process was how buildings should be done. And our manual calculations were confirming that the numbers were right: the shading was good, the loads on the perimeter were so low we couldn't believe it. It turned out to be a pretty successful project, a wonderful day-lit building.

A Rising Profile

The March 1998 cover of the *ASHRAE Journal* displayed the dramatic, multi-level concourse of Library Square, as designed by architect Moshe Safdie and featuring innovative underfloor air and ice storage systems created by KEEN Engineering. The *ASHRAE Journal* also proclaimed KEEN as the International Technical Excellence Award winner for the work done on the project, featuring an article authored by McCarry. KEEN designs were now gaining greater recognition from peers and they started to appear in other major trade and technical publications for the first time. "This was kind of a breakthrough," said Hydes. "Suddenly, we were being held as the leaders in North America of these underfloor air systems and what was being termed 'advanced office building' techniques. At this point, we were beginning to expand outside of our region. There was enough critical mass and confidence and knowledge of the experience we had gained in our home markets in Canada."

Through growing corporate and professional awareness, KEEN engineers were being hired to consult on these new approaches. McCarry flew across North America, offering his ideas and expertise, and influencing new buildings – many of which would include underfloor air distribution systems. KEEN staff also received invitations to speak to architects, consulting engineers and other professional groups throughout the Pacific Northwest. Increasingly, they were asked to participate in critiques and serve as judges for design competitions. The public speaking led to another new path for KEEN: giving technical presentations to international conferences such as the GLOBE Conference in Vancouver. There, Hydes, McCarry and others addressed their work on the Body Shop, C.K. Choi and Library Square projects – and for the first time started using the terms "high-tech path" and "low-tech path".

As a firm, KEEN was still exploring the boundaries of its low-tech and high-tech thinking about projects. Increasingly, KEEN tackled bigger buildings and faster schedules – often with a clear expectation on the part of clients for innovation. Where McCarry pursued techniques to make mechanical systems more efficient and smaller, Hydes focused more on designing buildings without systems. "In some ways, Blair and I had started a bit of an internal competition," said Hydes. "He's an amazing engineer, maybe the best in North America. Yet our approaches to design come from two completely different directions. So we'd challenge each other on our work – really strive for creativity, for innovation, and start a conscious dialogue to educate ourselves more."

Although KEEN was still predominately taking a traditional approach to systems design, they were listening to architects who expressed a growing interest in the ideas of "advanced office" buildings and then "green design." Internally, KEEN engineers began to compare notes more often about efficiency. That dialogue meant more emphasis on ventilation and daylighting, exploring thermal mass, and greater energy efficiencies. And above all, making designs as simple as possible.

Queens University Chemistry Lab

Location: Queens University, Kingston, Ontario

Type: Laboratory Research

Size: 140,000 sq.ft.

Completion: 2002

Architect: Brisbin Brook Beynon Architects

Sustainable Features:

- Heated by campus high pressure steam, perimeter ceiling mounted radiant panel
- Cooling provided by a stand alone cooling plant
- Underfloor displacement ventilation in lecture theatre with CO_2 sensor
- Strobic exhaust fans for fume hood
- Exposed ceilings in laboratory wing for thermal flywheel effect and to bounce daylight into interior spaces
- Thermal massing - exposed concrete in lab wing absorbs and releases heat to offset peak loads
- Full building control management
- Provision for future heat recovery on main air handling system

In early 1998, KEEN was retained for yet another high-profile, breakthrough project in Vancouver: the Telus/William Farrell Building revitalization. As western Canada's largest telecommunications company, Telus was the second largest provider of data, Internet, voice and wireless communications across the country. This $14 million project downtown would involve extensive interior and exterior renovations to the fifty-two year-old Robson Street wing of the building, resulting, eventually, in eight stories and 127,000 square feet of new office, retail/commercial and presentation space.

For KEEN, it would also be the third time they paired up with architect Peter Busby of Busby + Associates, following successes at the headquarters for the Association of Professional Engineers and Geoscientists of BC and the Revenue Canada office in Surrey. Busby had shown increasing interest in KEEN's recent work on high-tech buildings and its steady transformation toward more innovative designs. KEEN had plenty of experience with renovation and energy retrofits. But the Telus project would represent something new and completely different.

As originally constructed, the building's exterior walls consisted of brick and punched, single-glazed wood sash windows covered with adhesive reflective solar film. The client (Telus Accommodation Services) and the project team agreed on a number of green strategies to retain the entire building structure and re-use interior furnishings and fittings. In addition to seismic upgrading, Telus wanted the building to emphasize functionality, energy efficiency and an improved customer presence. Through its integrated design approach, the team recommended green features to include natural ventilation, natural daylighting, operable windows, and individual user-controller diffusers at each station.

Upon the project's completion in December 2000, the building became the first in Canada to incorporate a double-walled, triple-skinned exterior – receiving several major national and international design and construction awards. Another significant component of the project, the new cladding was designed to update the building's appearance, capitalize on its prime location and create a stronger Telus presence downtown.

The building's dramatic new façade surrounds the shell of the older office tower. Open and technically sophisticated, this new second layer was suspended 35 inches from the existing building face, providing an insulated perimeter plenum. The curtain wall, with its ceramic fritted pattern, allows a maximum amount of daylight into the interior (via lightshelves) while also shading the building. The envelope also features PV-powered ventilating fans. Electronic temperature sensors control the entire ventilation system to maintain temperatures inside the perimeter walls. Finally, the double-glazed curtain wall system has operable windows for additional ventilation.

KEEN's design input and energy modeling determined that this double-walled solution would optimize the natural ventilation in summer and act as a thermal buffer zone in winter. Ultimately, the building's exterior helped to achieve energy consumption levels that were 55 percent better than ASHRAE 90.1 and 35 percent more efficient than Vancouver's energy by-laws.

Within the building, KEEN's design scheme included some, by now, signature strategies. A raised access floor system was used to eliminate most of the conventional sheet metal duct work and provide flexibility for upgrading mechanical and electrical systems. Installation of the access floor was made possible by converting old equipment space with high floor heights in the building. Telus was also prepared to take advantage of the extra height by removing the existing suspended ceiling, thereby exposing the structure to provide dynamic thermal storage in place of a central chilled water system. In addition, existing steam heating was eliminated; instead, a new heat recovery system captures waste heat from large chillers in the adjacent telecommunications equipment building, therefore supplying nearly 85 percent of the Farrell Building's heating needs.

Back to the Choi

In May 1998, KEEN returned to the University of British Columbia campus in Vancouver to design an 18,000-square-foot companion to the C.K. Choi Building only forty yards away. UBC continued to pursue its goal of creating a more sustainable campus, with strong direction coming from the university president and director of planning.

This newest project involved the creation of a $5 million building to house the Liu Centre for the Study of Global Issues, established as a policy and conference center dedicated to critical cross-disciplinary studies to advise international policy-makers on issues such as global environment change, population growth and immigration. Arthur Erickson, Canada's preeminent practicing architect, and Stantec were commissioned to design the Center as a symbol of the spirit of globalization and a model of energy efficiency and sustainability.

Constructed on the site of Pan-Hellenic House and using the building's beams and other major components, the Liu Centre has two distinct elements: the one-story Seminar Wing with public spaces for meetings, receptions and conferences; and a three-story Research Wing for private offices, study, research rooms, library and boardroom.

Separate structural, mechanical and cladding systems responded to the different functions, topologies and environmental requirements of the two wings. "Breathing" glass façades face the surrounding forest, which provide cool, fresh air and views without the need for exterior sun shades or air conditioning. Narrow building widths allow for maximum daylight penetration and cross ventilation.

The project followed an integrated design process, with charrette participants – including a team of designers, users, operators and other stakeholders – defining a rigorous "green" agenda at the outset. Led by the university's Freda Pagani, they agreed on primarily a low-tech approach focused on building mass and orientation and energy efficiency that would utilize some of the same ideas and

BC Gas Operations Centre (Terasen Gas)

Location: Surrey, BC

Type: Post-Disaster Office Facility

Size: 180,000 sq.ft.

Completion: 2001

Architect: Musson Cattell Mackey Partnership

Rating: Commercial Building Incentive Program

Sustainable Features:

- Daylighting bounced deeper into occupied spaces with light shelves on south exposures
- Solar control for the windows is provided by sun shields
- Atria and operable windows promote natural ventilation
- Underfloor supply system provides displacement ventilation
- Exposed ceilings for thermal mass opportunities; night purging allows ventilation to cool the structure at night, resulting in reduced cooling load during the day
- On-site water management using bio-swales and streams
- Rainfall from the roof is directed into a holding pond for fire protection

both photographs Copyright © Jim Burns

Liu Centre for the Study of Global Issues

Location: Vancouver, BC – University of British Columbia

Type: Educational Center

Size: 18,800 sq.ft.

Completion: 2000

Awards: 2001 Consulting Engineers of British Columbia Award of Merit
1999 Association of Professional Engineers and Geoscientists of BC Environmental Award

Architect: Stantec Architecture in collaboration with Arthur Erickson

Sustainable Features:

- Natural Ventilation
- Thermal Mass
- Minimized energy usage to 50 percent better than code
- Use of storm rainwater for irrigation
- Low water consumption
- Re-use of waste heat from the campus steam trench for domestic hot water

techniques as the Choi building. Hydes remembered that design goals were frequently benchmarked against the Choi: "Let's be 20 percent better than Choi on energy, let's be thirty better on water. I thought we had a pretty naïve approach in some ways. In other ways, it was completely the right thing to do if we were going to keep progressing on sustainable design."

Despite the many similarities, the Liu Centre project had different site conditions and more design constraints than its predecessor. Specific mechanical system goals for the building were to minimize energy usage and maximize natural ventilation. Careful consideration was given to siting the building within the natural forest setting, with sensitivity to the quality of indoor light. KEEN designers discovered, for example, that pushing the building ten feet closer to the forest would have a profound impact on its performance: reducing internal temperatures by as much as 15°F.

KEEN's newly-acquired natural ventilation software also allowed designers to pose multiple "what if" scenarios for the project. It was equally important as a validation tool. "The software gave us much more data about where is the air flowing," said Hydes. "We still had some awkwardness in trying to reach forward in our designs, so this was a big investment for us. The first thing we did was a model of the Choi Building, which is interesting, because it did prove that ventilation rates were very close to the empirical methods we had been using. So it gave us confidence in both the empirical calculations and the software. With this new tool, we were able to inform the architecture and the architect."

In addition to natural ventilation, green techniques for the building included: dynamic thermal storage; a high performance low-e argon-filled glass curtain wall with operable windows and trickle vents; natural illumination complemented with efficient lighting fixtures on occupancy and daylight sensors; a heating system tied to centralized campus steam line with heat exchanger; and ultra low-flush toilets.

The Liu Centre project became yet another powerful learning experience for KEEN. Construction began in June 1999 and was completed in 2000. Ultimately, the building achieved similar results to the Choi in terms of the materials, as well as half the energy code requirements and approximately half the water consumption.

> The Telus project will set a new benchmark for environmentally responsible building design in downtown Vancouver.
>
> – Dr. Ray Cole, University of British Columbia, *The Globe and Mail*, August 17, 1999

Another breakthrough project for KEEN, the Telus/William Farrell Building's new double-walled, triple-skinned exterior contributed to energy performance significantly better than ASHRAE 90.1.

Green Myths, Green Challenge

In 1998, KEEN joined the Energy & Environmental Committee of the Architectural Institute of British Columbia (AIBC). There, KEEN's staff met some of the region's early thinkers in the green building movement, gaining more exposure to the projects and design philosophies that were moving the architectural community forward to a greener agenda. That exposure led to an invitation to KEEN to do a technical presentation on green engineering. On June 8, 1998, Hydes and McCarry presented the "Myths of Green Design" at an AIBC Professional Development seminar in Vancouver. For three hours, the two principals highlighted the importance of the "Integrated Project Team": committed owners, committed users, committed operators, in addition to the architects and engineers. They addressed the "Myths" as "Cost, Maintenance, Comfort, Ugly, Marketability, Risky, and Technology."

In the presentation, they also addressed the myth that green architecture had no place in Vancouver. A photo of Robson Square then flashed on the screen: the world famous, 1.3 million-square-foot development downtown designed by architect Arthur Erickson. Completed in 1979, the three-block complex featured an overarching space frame roof, a roof garden with ponds and waterfalls, a low-rise government office building, an indoor mall and outdoor plazas with hundreds of trees, shrubs and ground coverings. "Really, our goal for

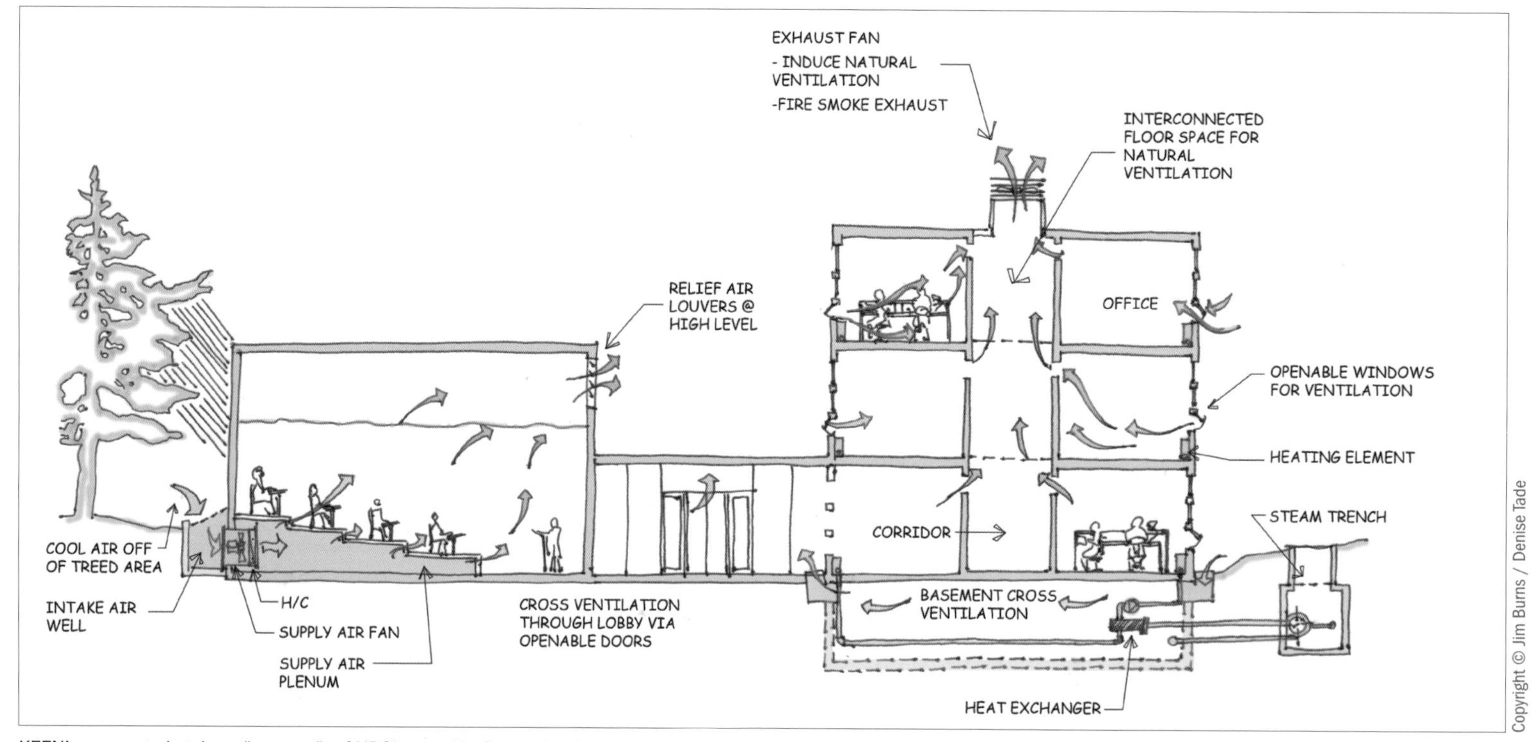

KEEN's concept sketch, or "cartoon", of UBC's new Liu Centre for the Study of Global Issues was used to illustrate all proposed heating, cooling and ventilation schemes at the design stage.

the presentation was to destroy these green myths," recalled Hydes. "Oddly enough, we didn't have any of our own truly green buildings to show. At that point, we were also starting to realize there's a whole world here that we had not been paying attention to."

In October that same year, the green building world came to Vancouver. The International Green Building Challenge (GBC) was a partnership of fourteen countries, focused on the development and testing of a new system of assessing the environmental performance of buildings. Ray Cole, of the University of British Columbia and a long-time advocate of green buildings, was the catalyst in bringing together designers from all over the world to this first-ever conference. Cole was also instrumental in the creation of the C.K. Choi Building.

During the course of the event, Hydes, McCarry and others listened carefully to detailed case studies presented on innovative green designs from Austria, Denmark, Germany, Japan, the Netherlands, Poland, Switzerland, and the U.K. The U.S. and Canada were as yet limited players in this international dialogue. But among those recognized for Canada were two recent KEEN projects: the headquarters for Revenue Canada and the C.K. Choi Building.

Going Greener

Meanwhile, 2,700 miles away on the shores of Lake Ontario, another milestone project for KEEN was underway. The Jackson-Triggs Estate Winery was expanding south of Niagara-on-the-Lake, with one hundred acres of prime land planted and a new building planned so the company could increase annual production to more than 100,000 cases of wine in time for the "crush" in the year 2000. The Niagara region of Ontario is the only one in eastern Canada with a mild enough climate to cultivate grapes. Jackson-Triggs, known for its fine Gewurztraminer and ice wines, had been named best Canadian wine producer at the 1997 International Wines and Spirits competition in London.

Don Triggs hired Kuwabara Payne McKenna Blumberg Architects, a progressive Toronto-based firm, to carry out his vision of a new winery that emphasized sustainable building features. The architect asked KEEN's Toronto office to join them in the venture. Project designers were given three environmental objectives for the two-story, 43,000-square-foot wine production and retail facility; it had to be agrarian in nature; building systems should have "conservationist" elements to reflect the wine-making process; and CO_2 emissions must be minimized.

KEEN's Toronto principal, Mark Mitchell, was intrigued about the project's potential. "We went there wondering how we could talk about sustainability with someone in the agriculture business, some who grows grapes for a living. We thought this could be very interesting."

Mitchell and his team also realized the project would present a perfect opportunity to utilize the green decision-making matrix again as they had for The Body Shop project. The original matrix had four primary components and a process that was often times complicated, requiring several hours to review each item with clients. This time, they elected to strip the format down into simpler decision-making points. "All those extra steps at the beginning of a project were designed to get in front of the client directly and clearly summarize the issues – that's where the decision-making matrix comes in," said Mitchell. "We make it up as if we're wearing the client's hat: 'I think I understand your business, so here's where the opportunities are.'"

Rather than attempt to determine specific paybacks and incentives, this time the KEEN matrix evaluated each strategy in two columns – showing a green value and a business value – with projected scores of A, B, C, D or E. The project team then rated every tactic according to its business or environmental benefits, such as the potential for saving money, or sustainability. The objective then was to match up the highest scores for both columns, making adjustments based on new assumptions, balancing choices, and other emergent criteria.

"After going through that exercise with the client, we'd look together for anything that changed," observed Mitchell. "For example, would we choose different strategies as a result of modifications to the decision-making matrix? So it was a very flexible tool. At the end of day, the client said, 'Let's proceed with those top six strategies.'"

As designed and constructed, the Jackson-Triggs Winery is divided into two principal components. The public space is located at the eastern end of the building and includes the tasting areas, a retail shop, and entertaining and administration areas. The western half houses the fermentation tanks as well as storage and barrel cellars, all of which are accessible to the public as part of the winery tour.

The Winery's green features include an on-site stormwater collection system that discharges to "soak away" pits. Daylighting is maximized to reduce reliance on artificial illumination systems. A 5 foot roof overhang on the south, west, and east faces minimizes direct solar heat gain and further reduces cooling energy requirements. Natural ventilation is maximized in all areas through the use of operable windows during shoulder seasons to reduce reliance on building heating and cooling systems. A displacement ventilation system is used for office, boardroom and lounge areas where high ceilings allow stratification and subsequent energy reductions and improvement in indoor air quality. And finally, the radiant floor heating system incorporates concrete mass for storage and a thermal flywheel effect.

In March 1999, KEEN again teamed with Peter Busby, now part of Busby Perkins & Will Architects of Vancouver, on a very different kind of educational facility: the Nicola Valley Institute of Technology (NVIT),

Copyright © Eduard Hueber / archphoto

Jackson-Triggs Estate Winery

Location: Niagara-on-the-Lake, Ontario

Type: Production and Retail Facility

Size: 43,000 sq.ft.

Completion: 2000

Architect: Kuwabara Payne McKenna Blumberg Architects

Awards: SB/GBC 2002 Canadian Presentation Project (Oslo, Norway)

Sustainable Features:

- On-site stormwater collection systems discharge to "soak away" pits
- Daylighting is maximized to reduce reliance on artificial illumination systems and reduces cooling loads and costs
- A 5-foot roof overhang on the south, west and east faces minimizes direct solar heat gain and reduces cooling energy requirements
- Natural ventilation is maximized in all areas through the use of operable windows during shoulder (spring and fall) seasons to reduce reliance on building heating and cooling systems
- A displacement ventilation system is used in office, boardroom and lounge areas where high ceilings allow stratification, subsequent energy reductions and improvements in indoor air quality

University College of the Cariboo. Located in Merritt, in the interior of British Columbia, NVIT was one of Canada's first post-secondary facilities shared by a native and non-native institute. *Canadian Architect* magazine noted later that the "First Nations learning facility embodies the exploration of innovative technology alongside the traditions of a burgeoning aboriginal revival in British Columbia."

Phase 1 of NVIT's building program would include classrooms, faculty offices, social spaces, labs, a bookstore, cafeteria and library arranged along an interior street. At nearly 74,632 square feet, this first building on campus was tucked into a forested, south-facing slope on the outskirts of Merritt. It emerges from the hillside as a two, then a three-story structure.

Inspired by traditional native structures and elements of aboriginal design in the region, the building's semi-circular shape is a meaningful and recurring native theme. Like other contemporary First Nations projects in Canada, NVIT's arbor and building are oriented on the cardinal points of the compass, with the main entrance facing east and a non-hierarchical arrangement of functional spaces. Busby and KEEN chose the building form, orientation and materials primarily in response to environmental and micro-climatic conditions, particularly to take advantage of the ample sunlight available throughout the year in Merritt.

Again, natural ventilation was a central element of KEEN's design. The main part of the building features a glazed ventilation stack with operable windows located along the perimeter spaces. An atrium within the center of the Institute creates the pressure differential necessary to naturally ventilate the building, while motorized windows at the top of the atrium allow warmer air to escape.

Earth-sheltering further enhances the thermal insulation of the building envelope. Heat gains stored by the concrete structure during the day permit excess heat to be released through natural ventilation during the cooler evening hours. A portion of the roof is planted with an indigenous shrub to further assist the thermal performance of the structure and tie the building visually to its surroundings. KEEN innovations also included use of the cladding module to incorporate horizontal bands of PVC-framed tilt-and-turn windows, shaded by moveable panels of yellow cedar sun-shading louvres. In addition to natural ventilation and dynamic thermal storage, the building design included high efficiency condensing boilers, an air-cooled chiller and high efficiency lighting systems – allowing for a larger thermal comfort range and achieving substantial energy savings.

Passing the Torch

As mid-year 1999 approached, KEEN appeared ready for its next big step as an engineering firm. Tom Johnston, who had spent nearly twenty years building an organization of professionals, had already started his retirement process. He understood the company needed to move forward and could capitalize on these new high-tech and low-tech designs – although he still was not convinced that sustainable projects could be compatible with the traditional business end of mechanical engineering. Yet Johnston, who had taught people how to be "keen" – sharp, enthusiastic, focused – knew it was time for somebody else to take the lead. He remembered the example set by Jim Keen, the founder and innovator, who had passed away on January 14, 1999.

"Tom had talked to most of us about his eventual retirement, and he was pretty good at apprising people of what's coming down the pipe," said Mitchell. "But he was also unclear about what he wanted and how it could be done." So Johnston brought in a respected ownership transition adviser – who concluded the future of KEEN was 'M^2HA': Mitchell, McCarry, Hydes, and Paul Anseeuw, who then managed the Victoria, BC office. He also recommended that Johnston stay with KEEN as chairman of the board during the transition period. Ultimately, Johnston agreed to a buy-out package and the new senior managers, M^2HA, agreed to become equal partners. McCarry would continue to focus on technology firm-wide. Mitchell would continue to expand KEEN's presence in Toronto and eastern Canada. Hydes would move to Seattle to open a new office there. And someone needed to step into the president's role.

"We were all quite different and had all done quite different things," recalled Mitchell. "My own research on ownership transition was that we needed the president to be the guy with the most personal connections to the most number of senior people within the company. And that had really been Kevin up to that point." In June 1999, forty-two year-old Kevin R. Hydes – a native of Leeds, England and graduate of Newcastle Polytechnic who had emigrated to Canada in 1982 – was appointed KEEN's third president.

Hydes remembered, "We had just finished a whole series of breakthrough buildings at this point, all attracting a number of visitors. Engineers and architects wanted to see them because they were two or three years ahead of most of the other work in the region. So from the outset, my vision for the firm was really clear. We could seize the opportunity to become a pioneer in the field of green engineering in North America, not just a parochial firm serving one market. I was completely convinced that this was the direction we had to go down. We just needed to decide we wanted to do it."

At KEEN's next company-wide management conference, Hydes invited several architects as guest speakers and challenged his co-workers to pursue a greener agenda:

> Our speakers put this amazing picture together of why they saw green architecture as the future of their work. When I said 'We are a green firm,' I think everybody knew I was serious. It was clear then that KEEN was looking for something to grab a hold of, and there were definitely people already aligned with the idea. At the same time, I didn't actually take over the position with a promise to take us on this green journey. Becoming president and launching this green agenda were two separate, conscious actions. But once we agreed where we wanted to be in five years, we really did set our course, a new direction.

Living the vision and getting alignment across the company and its multiple offices remained a challenge, however. KEEN's reputation as underfloor air or even sustainable engineers continued to grow – as did the number of new clients and projects calling for their unique areas of expertise. The momentum internally was growing as well, as more and more staff expressed interest in these greener techniques and design strategies. Yet that green momentum was not consistent across the company nor among the principals at every one of KEEN's ten offices.

Hydes explained further:

> We were certainly not aligned as a group at that point. A pretty common counterview from another principal would have been, 'I like what you're saying, Kevin, I actually believe in what you're saying, but my clients don't want it in my market.' Some believed that green engineering represented nothing more than just a smallish portion of the market and was probably only a West Coast phenomenon at that. I think what started to break the ice was that architects and clients in a given region started to call the KEENers in the nearest office. So slowly and one by one, our regional offices started to get projects with green objectives. Each new client was demonstrating the unexpected level of interest out there.

Then came what Hydes referred to as KEEN's "stumbling-around phase", a time when principals and associates struggled to gain the necessary technical understanding of this new design thinking and application – and keep it profitable and share what they were learning with others in the company.

> We knew where we wanted to go, the direction had been set, but there was that stumbling around you'd expect to see at first, where an office without a lot of experience in certain techniques would 'stub their toes,' just like I did five years earlier with the Choi. We were changing the firm at that point, so we went through quite a bit of stumbling around, struggling to get our internal systems in place. One person would be designing a natural ventilation

British Columbia's Nicola Valley Institute of Technology reflects cultural characteristics of First Nations students, achieves state-of-the-art learning spaces and presents a responsible effort to lower energy consumption.

Nicola Valley Institute of Technology

Location: Merritt, BC

Type: Post-Secondary Facility

Size: 74,632 sq. ft.

Completion: 2001

Awards:
2002 Vancouver Regional Construction Association, Awards of Excellence, General Contractors Award of Merit
2002 The Wood Design Awards, Citation Award
2002 Lieutenant Governor of BC, Medal for Excellence
2002 International Green Building Challenge

Architect: Busby Perkins + Will

Sustainable Features:

- Natural vegetation and window shading to minimize direct solar gain
- Natural ventilation (operable windows), dynamic thermal storage, high efficiency boilers, and high efficiency lighting systems result in energy consumption 30 percent lower than the National Energy Code
- User-controlled, operable windows, relief grilles between rooms and atrium and sensored/motorized windows in the atrium. Motorized windows at the top of the atrium allow warmer air to escape
- Thermal chimney provided by atrium reduces cooling loads during shoulder seasons
- 35 percent more efficient than ASHRAE 90.1
- Unit ventilators with heat recovery provide heating and cooling to classrooms
- Heated water and chilled water are distributed via two-pipe system

building without knowing about a similar recent project done by someone else. The two engineers didn't talk, so we might have two different solutions to a similar problem in similar conditions. So we were realizing how critical it was to do a better job of communicating technical issues around the company.

High Performance Energy

Hydes was convinced that for KEEN to be a green engineering leader, it also needed to become a multi-disciplined firm – and that meant expanding into both electrical and energy work. Some of the energy foundation was already in place. In 1996, KEEN's Edmonton office had formed an important alliance with Rose Technology Group, one of North America's largest energy performance contractors. Rose focused on capturing energy savings on behalf of building owners through better energy performance. Essentially, an owner could use those future savings to defray capital costs upfront; then over a period of seven, eight or even ten years, the costs were re-paid through the additional energy savings.

The KEEN/Rose partnership was designed to help Rose expand its operations to western Canada, while KEEN could offer clients new services such as energy project identification, commissioning, monitoring and verification, energy audits, energy modeling and more. KEEN provided the engineering expertise to support it, while Rose contributed its high level of energy expertise. Chris Jepson, KEEN's principal-in-charge of the Edmonton office, took responsibility for the new venture.

Staff were added as business developed, and KEEN/Rose had a number of successes in Alberta where, said Jepson, "the economy was freer and prices went where they needed to go." But the market for energy performance never quite took hold in British Columbia's more regulated environment, which had more government and utility restrictions on infrastructure and energy upgrades. "What we learned," Jepson said, "was that energy performance projects took a while – anywhere from one to two years – from initial contact to where any actual work occurred. So you had to allow it to grow slowly, and you had to find the right people to help sell it. Even after a couple of years, we still had a viable business and believed in it."

In late 1999, Rose was acquired by Cinergy, the Cincinnati, Ohio-based utility and renamed Vestar Energy Group to offer facility and infrastructure solutions to its institutional, commercial and industrial customers throughout North America. As a result, KEEN chose to form its own team dedicated to supporting clients with consulting services that were not tied strictly to a performance contractor. And the KEEN Energy Solutions Group was born.

"KEEN was looking for a way to help our clients deal with their energy challenges," said Jepson. "In a way, it took our blinders off. The new group allowed us to look at a far wider set of opportunities than just buildings, as well as our involvement in the integrated design process."

Beyond Alberta, Energy Solutions started to gain new clients in British Columbia, Ontario, and other provinces. Several senior managers with Rose joined KEEN. And a new partnership with Cinergy was underway – with Cinergy funding new energy performance projects and KEEN providing the expertise on energy savings in buildings.

Over the Border and Through the Woods

By now, the U.S. Green Building Council (USGBC) had launched the pilot phase of its new LEED® (Leadership in Energy and Environmental Design) Green Building Rating System, a voluntary, consensus-based national standard for developing high-performance, sustainable buildings. Hydes immediately recognized the potential for transforming the building market. LEED® was created to promote integrated design practices and stimulate green competition. And just as significant, it would offer training, project certification and professional accreditation.

Hydes was introduced to Lynne Barker of Seattle, a USGBC board member who worked for Sellen Construction Company, one of the largest contractors in the State of Washington. In 1998, the City of Seattle had developed a set of recommendations to develop a green building standard and promote industry education. Seattle then received funding to extend those recommendations throughout the Pacific Northwest bio-region – to Washington State, Oregon and as far north as Vancouver, British Columbia. Over the next year, more than two hundred industry professionals participated in a series of workshops, culminating in a report, the "Northwest Regional Sustainable Building Action Plan", released in March 1999.

That report led to Seattle adopting the first green building policy in the nation, requiring all city-funded projects to achieve LEED® Silver certification. Barker invited Hydes and others such as architect Peter Busby to meet regularly with the local development community to share their ideas on green construction.

Barker recalled:

> We were trying to encourage our customers to embrace green building. In the City of Seattle, design-build was the conventional approach, so the business model didn't really support green engineering, because mechanical fees were based on equipment installed in the building. There was no incentive to design equipment out of a building. So every chance I got, I would get this idea in front of people – because KEEN had the commitment and the resources as a design team to work on these kinds of projects.

Seattle would also be the location for one of the first-ever LEED® workshops at this time and KEEN was asked to be a sponsor. On that day, the blue KEEN logo was displayed prominently on the roster as the event's "Platinum" sponsor – alongside logos for Turner Construction and HOK. "We had never sponsored something like this before and didn't know what to expect," said Hydes. "To be honest, we didn't really know what we were doing. We had no plan. But it seemed we were instantly in the game, right in the middle of this circuit. Again, I definitely felt like the green direction was what we had to do."

As president, Hydes now started to spend more time in Seattle and Portland, cities still recovering from a recent recession but where interest and funding of green projects was accelerating. In particular, the Pacific Northwest's design community was eager to learn from KEEN – and invitations to speak and to participate in new projects soon followed. Among those interested in KEEN's green design ideas and experience was Bert Gregory of Mithun Architects + Designers + Planners in Seattle. Mithun and KEEN soon teamed up to create the new Puget Sound Environmental Learning Center for the Brainerd Foundation. Later renamed "IslandWood", this unique educational facility for school children would be located on 250 acres on the south end of Bainbridge Island in suburban Seattle.

Founders Paul and Debbi Brainerd envisioned the center as a place "to inspire environmental and community stewardship by providing hands-on learning experiences that link science, technology, and the arts in a natural setting." The curriculum was designed to teach students about sustainable lifestyle choices, in turn, providing the 4th and 5th graders with a much deeper understanding of their influence on nature. It was also important to the Brainerds that exceptional environmental standards be met in building this teaching environment. At IslandWood's heart is a campus of sustainably-designed buildings: an interpretative center, learning studios, dining area and kitchen, administrative offices, arts studio and maintenance building. In planning the site and buildings, the project team and clients agreed their goal would be "gold" – the new LEED® Gold criteria.

The team employed "eco-mapping" to locate appropriate building sites and energy modeling to determine optimal building orientation, shape and configuration. KEEN's innovative mechanical design took full advantage of the area's mild climate, strong cooling breezes and the forest's natural shading to provide ideal conditions and profit on savings in energy and capital costs. In targeting LEED® Gold, KEEN relied heavily on natural ventilation as well as active and passive solar schemes for hot water to power 50 percent of classroom energy needs in IslandWood's Learning Studios. Primary campus buildings also included high-efficiency, in-floor hydronic heating. In addition to a rainwater collection system at IslandWood, all wastewater is fully recycled and collected for re-use via a "Living Machine™" and a subsurface flow constructed wetland.

IslandWood would serve as an important learning experience for KEEN Engineering. After the firm's initial LEED® submission, the USGBC challenged their findings on eleven points. In response, KEEN provided supporting material including thermal modeling reports as proof the project met the LEED® criteria. IslandWood then became the first LEED® project constructed in Washington State, the third in the United States and fourth in the world (see Chapter 10).

In September 1999, KEEN was invited by a second Seattle firm, Mahlum Architects, to consult on another progressive educational facility. Located just southwest of Seattle in Olympia, Evergreen State College had a national reputation for its innovative, interdisciplinary, collaborative and team-taught academic programs. Known too, for its natural setting in the woods, the college was committed to sustainability and required eco-friendly designs for all new building construction. The new "Seminar II Building" would house the college's Public Service Center, Evening and Weekend Studies Program and a small satellite café. Mahlum knew of KEEN's work on the C.K. Choi Building and needed an engineering consultant on the project team. The architect's initial concepts illustrated a logical design approach to a campus building: organize it into three stories, with lecture theaters on the lower level, computer labs in the middle, and academic offices on top.

Mahlum Architects requested KEEN's input and they responded with a completely different solution: three separate buildings clustered together. They reasoned that comfort issues in the theaters would be completely different from those in the computer labs, and yet again compared to those in the offices. "We were starting to think more about a building and its component parts in terms of an energy budget," said Hydes. "It was our belief that things like corridors and staircases and atriums shouldn't use any energy – they're just transient spaces." Instead, KEEN's diagrams essentially de-coupled the original building concept, creating a completely different architecture that downscaled, in fact, eliminated the entire mechanical system. The new designs showed the building as naturally ventilated. And placing the lecture theaters in the ground would allow cooler temperatures in summer and enough ground tempering and body heat for people to stay warm in the winter – and no supplemental heating system was necessary.

The architects were very receptive to the ideas. Again, Hydes had relied on empirical data and historic precedence, saying, "This is what's going to happen in these spaces. We've seen these buildings with good daylight for hundreds of years. With the right proportions, we know those buildings actually work – they're quite comfortable. You're going to get temperatures that are clearly higher if you air condition the building, but this design just eliminated a million dollar system. So we think it's still worth doing in this climate and location."

"But nobody's going to trust you based on an intuitive diagram," he added. "We knew our methodology had to incorporate analysis and modeling to validate our recommendations." KEEN used its Thermal Analysis Simulation (TAS) software to predict the thermal response of individual naturally-ventilated zones. Additional calculations and modeling demonstrated how man-made shading devices and placement of the office modules closer to trees could minimize solar gain, resulting in still more energy savings. The project team and clients were convinced.

Meanwhile, Back in Paradise

In a land of glacial lakes, hot springs, sand dunes and forested slopes on the eastern edge of Jasper National Park, the Town of Hinton,

Alberta lies at the heart of a recreational paradise in the Canadian Rockies. Hang gliding, skiing, and hiking trails – Hinton (population 10,000) is a four-season playground for tourists. It is also the center of several important resource industries, such as modern coal mines and one of the most profitable pulp mills in Canada.

On November 26, 1998, Hinton's town hall was completely damaged by fire. Enter Vivian Manasc. As senior principal of Manasc Isaac Architects, one of Edmonton's largest architectural practices, she and her firm were known for their award-winning work on municipal buildings and schools – as well as their growing commitment to sustainable building design. The Hinton government wanted Manasc, and Manasc wanted KEEN as part of the fully integrated design process.

The new 30,000-square-foot Hinton Government Centre would house municipal and provincial offices when completed – and construction had to be fast tracked. In this resource-rich region with a booming tourist industry, the town was conscious of minimizing the project's impact on the surrounding environment and the importance of making a statement about energy efficiency. The new building qualified for a $70,000 federal grant recognizing projects that incorporate significant energy-saving measures.

First, the design team modeled building orientation, to take advantage of an east-west exposure as well as a dramatic view of the mountains. The building is narrow to give as many employees as possible access to natural light and reduce the size and use of lighting systems. The three-story structure is built into a wooded hillside and takes advantage of natural drainage, offsetting the need to move water into the town's storm sewer. Clearly, Hinton Government Centre represented a breakthrough project for KEEN's Edmonton office. "It's an attractive building, it speaks to what you can do," said Chris Jepson. "It really is a building that you enjoy going to. It's simple." And Manasc noted in the *Edmonton Journal*: "We're on the leading edge of a new movement integrating building design, making healthy environments, smart buildings, places people can actually live in and work in."[2]

Evergreen State College

Location: Olympia, Washington

Type: Post-Secondary Educational Facility

Size: 150,000 sq.ft.

Completion: 2004

Awards: AIA/COTE Top Ten Green Projects, 2005

Architect: Mahlum Architects

Sustainable Features:

- Maximum daylighting and minimal mechanical loads
- Natural ventilation with consideration being given to comfort, climate and sustainability
- Detailed analysis, modeling and the use of both natural and man-made shading devices incorporated in the early design process
- Thermal Analysis Simulation (TAS) used to predict the thermal response of individual naturally ventilated zone

Located in the Canadian Rockies, the Hinton, Alberta Town Hall was re-built after a major fire – into a new 30,000-square-foot structure that takes full advantage of daylight, natural drainage, natural ventilation, and water-source cooling.

The upper two levels of the building include office space and a central atrium, allowing warm air to rise in the summer and be evacuated through operable windows. In fact, all windows open in the building to maximize natural ventilation according to the season. Municipal office employees can also control their space by adjusting air flow through workstation vents connected to an underfloor air distribution system. The provincial government offices chose, instead, to specify a conventional supply air system suspended from the ceiling.

The Town Hall's building design also features a critical shading scheme. It is essentially a wood building, and all locally harvested. Yet, eleven out of the site's fifteen acres were left undisturbed, still covered in coniferous trees to maximize natural shading and buffer the building from western winds. Much of the exterior features solar shades, together with light shelves on the inside to capture as much daylight as possible and minimize energy usage.

KEEN made another important discovery about the site that would have a far-reaching impact on the project's success. The main water supply line for the town ran right past the Town Hall, estimated at an average daily flow of about 2,300 gallons and a temperature of around 43°F. KEEN's designers saw it as a natural source of cooling and performed the calculations. They were able to show that even with the building's peak cooling demand – and ignoring any beneficial impacts of natural ventilation or the underfloor air system – Hinton's water supply could absorb the building cooling load, and the system would see only a slight increase of about 34°F in water temperature.

"As a consequence of our cooling solution, together with the upgraded envelope, lack of cooling towers and chillers, and low energy consumption on lights, we were able to reduce energy use by more than 50 percent compared to a conventional new building," said Jepson. "In a climate where the temperatures can fluctuate from -40°F to +90°F, you can get some wide variations. The Hinton Town Hall is not complicated – and the owners embraced it." Hinton Town Hall was completed in 2001.

Five-Year Milestone

Entering 2000, KEEN's new partners looked back on the past five years – since Library Square, since the C.K. Choi Building and Revenue Canada – and saw a clear path, still forming, toward becoming a green engineering firm. By then, KEEN was attending new conferences, meeting new people – and attracting a very different type of client. "We were intentionally saying that these green projects are what we're trying to focus on now," said Hydes. "We were involved suddenly with a new group of thinkers and doing all these cool buildings. As a firm, we were definitely becoming kind of green, but there were still two teams: the low-tech and the high-tech. So we had to work on that."

IslandWood

Location: Bainbridge Island

Type: Educational Facility

Size: 70,574 sq.ft.

Completion: 2002

Rating: LEED® Gold

Architect: Mithun Architects + Designers + Planners

Sustainable Features:

- On-site treatment systems created for all wastewater, including a Living Machine™ to treat greywater and blackwater for flushing toilets and irrigation; Constructed wetlands treat the wastewater at the lodges and graduate student cabins
- Solar hot water used for dining hall and three lodges
- Solar heat for heat-dependent buildings oriented east-to-west for maximum solar gain, with overhangs for south-side protection
- Photovoltaic system – a 23 kW PV array powers 50 percent of the classroom energy needs at the Learning Studio building; small PV panels power building exhaust fans
- IAQ: CO_2 monitors in primary spaces monitor air quality
- TAS modeling used to locate windows and shape building for optimal energy performance
- Primary buildings heated with high-efficiency in-floor hydronic heating
- Natural ventilation
- Partial rainwater collection system at educational buildings used for student education, boot washing and local irrigation
- Low-flow fixtures to conserve or eliminate water use

McCarry, Hydes and other designers were being asked to consult more and more on projects in Portland, Seattle and the East Coast. They were also increasingly aware that their approaches and their design techniques on these new kinds of buildings were distinct and completely unique. "This is when we realized that KEEN was definitely still going down these two separate tracks, that we really needed to combine both of them," said Hydes. "No building should be strictly high-tech or low-tech; it should be whatever it needs to be." At subsequent partners' meetings, they started to share more information on various projects underway. Often, they would hear of KEEN's own projects from other engineers or architects in their region, or happen to see a technical article or conference presentation. Hydes suggested they compile everything into one document, in effect, creating a flowchart with two columns that list each of the firm's high-tech and low-tech projects. It became KEEN's "Family Tree."

Said Hydes:

> We realized it was a bigger body of work than we'd thought, and it could be traced back to these two seeds, high-tech and low-tech, that had been planted. We had not been consciously managing the process, it just happened: I might have worked with an engineer on one project; then that engineer would work with a principal on another project – and they'd take some of those ideas to another piece of work. So that created a new branch coming out of the big family tree.

As a result, KEEN began to bridge the two design approaches by establishing formal internal communications systems, so project engineers could learn about specific innovations on projects that were completed or still in progress. Likewise, they were enthusiastic to post project data to share with their colleagues and other offices as well. "Once we started to exchange information among designers, we were able to package our thinking much more clearly as a firm at this five-year mark," added Hydes.

Among the World's Best

For many at KEEN, the fusion of high-tech and low-tech thinking began to merge visibly in the firm's newest project for York University. The Computer Science Building on the university's campus north of Toronto would combine displacement ventilation and stratification as well as passive heat reclaim and passive preconditioning of supply air from an underground airflow plenum. Most importantly, the building's final design involved natural ventilation, a serious challenge in Toronto's cold climate.

As designed by Busby + Associates and Architects Alliance, the new 115,000-square-foot, four-level building included computer labs/ classrooms, lecture theaters and offices connected through two large atrium areas, which had a central role in providing fresh air and daylight throughout the facility. The atria, in conjunction with a thermally efficient building form and orientation, operable windows and thermal ventilation stacks, resulted in energy efficiency that exceeded ASHRAE's 90.1 standard by 40 percent.

That level of performance and the building's innovative design led *World Architecture* magazine to name it one of the best green buildings of 2002. The publication noted:

> There is nothing especially novel in the way York University's Computer Science Building handles green design – no dazzling gizmos or weird new materials – but the three-story courtyard building is a lesson in putting sensible environmental thinking at the heart of a scheme.

KEEN's new, combined design approach also included measures such as distributed fan coils for peak summer and winter loads and natural ventilation and/or mechanically-assisted ventilation during the shoulder seasons. Innovative use of millwork as a work surface, supply air plenum and electrical chase replaced the traditional supply air ductwork. The building exterior featured louvres along the principal façade to prevent the summer sun from driving heat into the central atrium. Walls were highly insulated. And a sod roof was used to collect runoff for irrigation purposes and reduce the peak load on the stormwater system.

Lorna R. Marsden, York University's president, later expressed her thanks at the building's dedication in August 2001: "This beautiful and environmentally-friendly building has been needed for some time by our large and successful faculty of Pure and Applied Science. Opening this building now provides space for some of the new students we expect to select York's science faculty in the coming years."

Green Declaration

In May 2000, KEEN Engineering officially returned to the U.S. with the opening of its newest office in Seattle. Shortly afterwards, KEEN entered several buildings in the 2000 International Green Building Challenge, scheduled for Maastricht, The Netherlands in the fall. Ultimately, two of three projects selected by the Canadian GBC 2000 National Team for the competition were KEEN's: the Telus/ William Farrell Building Upgrade and the York University Computer Science Building. Two additional KEEN buildings were selected for poster presentations: the BC Gas Operations Centre and the Liu Centre.

If there was ever a formal declaration of KEEN's new direction as a firm, it came at the 2000 Fall Management Conference in Calgary. Partners, associates, and staff were interested in learning about KEEN's new vision going forward – its new corporate message to the building industry at large.

Named one of *World Architecture's* best green buildings of 2002, the York University Computer Science Building near Toronto utilizes its "Tree Atrium" for fresh air and daylight throughout the three-story office space.

Just completing his first year as KEEN's president, Hydes remembered the meeting clearly:

> We were definitely a good firm, but it seemed like we needed a goal that was both symbolic and literal. So as a group we agreed we should do just two things: aim to be the number one green engineering firm in North America and become the only truly national Canadian mechanical and electrical firm. So here we were, a forty year-old engineering firm, deciding to head down two completely new paths – simultaneously – and with intent to double our size in the next five years. I think it did give people a very clear idea of where we wanted to go. These two strategies were aligned to green thinking and to each other.

KEEN Engineering was now poised on the brink – of green.

Being KEEN Green

Transforming the Industry

...it has the quiet ping of truth, like a single, clear, perfectly struck note hanging in the air in the hushed silence of a full auditorium at the end of a quiet movement of a Mozart piano concerto. There is no need to say much of anything; the quiet truth speaks for itself.

– Jim Collins, *Good to Great: Why Some Companies Make the Leap...and Others Don't*

For two years, KEEN had pushed itself to be an integrated firm, to perform at a continuously higher level of sustainable design, and to be green leaders. By now, the changes at KEEN were becoming just as profound internally – in the firm's new design philosophies, processes and expertise – as they were externally, through a growing portfolio of green projects and newfound ability to influence clients, peers and the industry.

"To a large extent, we can attribute much of that to our internal desire to change behaviors and habits – the way we think and design and deliver," said Hydes. "But then again, it was our clients who sometimes dragged us to where they wanted us to be, who told us they wanted something different. In fact, they were the ones who put a stake out there somewhere that we'd never seen before."

KEEN Strikes Gold

To the west of Vancouver, one-and-a-half hours away by ferry, is Victoria – the capital of British Columbia. And just outside of the city in the community of Saanich, surrounded by farmland, is a nondescript industrial development that would become a landmark project for KEEN as the first LEED® Gold-certified building in Canada and the third LEED® Gold in all of North America. All because of the client.

"Vancouver Island Technology Park" began as a feasibility study by its owner and property manager, the British Columbia Buildings Corporation (BCBC), which planned to turn the former Glendale Hospital into a new office complex. Two years after that initial study, in August 2000, Idealink Architecture and KEEN's Victoria office were contacted to move forward once again with the conversion project.

Phase 1 of the redevelopment by BCBC would be geared for small-to-medium sized companies needing between 2,000 and 12,500 square feet of space. Work proceeded to renovate the existing 165,000 square foot, three-building complex into a technology park at a projected cost of $11.9 million. Demolition, for example, included tearing out ten-foot wide exterior walkways around the perimeters; and it turned out the demolition contractor was quite entrepreneurial, selling most of the salvage elsewhere so that 95 percent of the materials were either recycled or reused.

Construction was well underway when the project was halted suddenly. Joe Van Belleghem of BCBC stepped forward, declaring that he wanted Gold – LEED® Gold – for the new building to reflect the new mandate of BCBC for greener buildings. Change orders were issued. And the project team was called together. Tom Wilson, manager of the Victoria office for KEEN, recalled: "He wanted a Gold building, essentially the first LEED® Gold building in Canada, saying, 'you boys need to help me get there some way or another.' But they didn't talk first with our green knowledge leaders in Vancouver – they didn't know them. This was a project involving our Victoria office, a relatively small office of eleven engineers."

With the project again underway, KEEN's engineers recommended several green strategies for the thirty-five acre site and the existing two- and three-story structures. A common sewage and stormwater collection system had served the former hospital so KEEN created a highly innovative stormwater management plan using a system of onsite grass swales, and treatment and retention ponds. The new grass and gravel pave system also helps to manage stormwater flows. Engineers specified water-conserving measures such as waterless urinals and dual-flush toilets using stored rainwater.

The hospital's original 100 percent outside air systems were much bigger than necessary for an office environment, so KEEN added variable speed drives and has since added some return air capability. Additional ventilation was supplied by a single large air handler in each of the two T-shaped buildings (on the perimeter of the complex) and two smaller air handlers in the center building, one of them multi-zoned to service about half of the space. For the renovation, KEEN had the multi-zone unit refurbished, rebuilt all controls to DDC, and replaced the large air handlers with 100 percent water-loop heat pump systems to condition the T-shaped buildings.

"I think that this is a good example that green doesn't have to look flashy," said Wilson. "Green can just be well thought out. You don't have to have an architectural monument to get the job done right and still be effective. As far as I'm concerned, the building is a shining example of how sustainable design can be mainstream. There it is."

By 2005, all tenant spaces were occupied or under construction, and BCBC is considering plans for additional construction on site. As a result of the Vancouver Island Technology Park (VITP), bus routes have changed, bike racks and a network of cycling trails have been added to encourage cycling as an alternative mode of commuting, and the city has begun mandating LEED® Gold level projects for every municipal building. In addition, the BCBC began using this project as a case study in developing LEED® certification for broader use in British Columbia.

For Tom Wilson, VITP was a significant learning experience:

> I went through a bit of a transition as a result of this project. And it's not because the mechanical systems are necessarily all that stretched; it's difficult sometimes for technical guys to think outside the box. You know, the box is pretty safe, and we're pretty familiar with all the bits that are inside the box. There's a level of comfort there – and pushing the envelope and doing a green thing. This project opened my eyes to...just how exactly do you think outside the box?

KEEN would soon strike gold again – this time on the White Rock Operations Centre, the second certified LEED® Gold certified project in Canada and the country's first Gold for New Construction.

Just seven months after beginning VITP, KEEN's Vancouver office was teamed with Peter Busby on a municipal building project for the City of White Rock, British Columbia, an oceanside community along the coast south of Surrey and Vancouver. In March 2001, the City of White Rock appointed Busby and his team to design its new operations building. Located in a predominantly single-family residential neighborhood, the center would serve primarily as a base for civic utility vehicles and street-washing vehicles, with offices for nine full-time staff and eighteen field staff who use the building two to three hours a day.

The purpose of this small facility may not have been remarkable. But upon its completion two years later, it became extraordinary for its use of light and water and innovative use of solar and other renewable energy strategies. The building's reduction in energy consumption is projected to be 55 percent more efficient than the ASHRAE- 90.1 standard for energy-efficient buildings in North America.

White Rock's mandate was to make the new building as environmentally sustainable as possible, in accordance with the city's own policy of promoting green strategies in all municipal developments and planning initiatives. As the project evolved, a decision was made to pursue LEED® certification.

Designed as a single structure with two separate pavilions, the Operations Centre was constructed atop the site's original operations building. In fact, Busby took advantage of the site's abandoned wastewater treatment plant, using its concrete clarifiers and pump house as a foundation for the new facility. The 6,550-square-foot building effectively combined a variety of green strategies to satisfy the LEED® rating system. These strategies include: operable windows, with cross-ventilation supported by roof overhangs; a wall trellis and sunshades to eliminate the need for air conditioning; extensive daylighting; a green roof to reduce heat gain; and solar hot water tubes. The city also purchased green power certificates from BC Hydro

Vancouver Island Technology Park

Location: Victoria, BC

Type: Three Office Building Complex

Size: 165,000 sq.ft of renovated space; 235,000 sq.ft. of new office buildings

Completion: 2002

Rating: LEED® Gold V.2.0

Architect: Bunting Coady Architects & Idealink Architecture

Sustainable Features:
- Purchased Green Power
- Materials selected with low or no VOCs in order to maintain good indoor air quality and occupant comfort
- 100 percent stormwater treatment and infiltration onsite
- Use of recycled and reused materials, construction waste diverted
- 99 percent of all deconstruction and reconstruction waste salvaged or sold
- High performance envelopes and systems showcase building integrated photovoltaic panels

White Rock Operations Building

Location: White Rock, BC

Type: Operations Building

Size: 6,545 sq.ft.

Completion: May 2003

Rating: LEED® Gold

Awards: Consulting Engineers of BC Award of Merit 2004

Architect: Busby Perkins + Will

Sustainable Features:

- Commitment to purchase Green Power Certificates
- Solar electric power generated onsite
- High efficiency fixtures and mechanical system reduce energy consumption
- Indoor air quality is enhanced with natural ventilation, operable windows, and direct ventilation in high contaminant areas to reduce pollutants
- Daylighting is a main feature in the building with sunshades, wall trellis, overhangs, and light shelves to optimize lighting and control heat gain
- Reduction in site water use by approximately 90 percent (2 million liters/year)
- Reduction in building water usage by over 20 percent
- Reduction in energy consumption by about 55 percent over ASHRAE 90.1

Photo Courtesy of Busby Perkins + Will

Copyright © Jim Burns

to augment its on-site photovoltaics to support a low-energy heat pump. Yet, perhaps the most distinguishing element of the new Centre involved water, given that the city of White Rock receives forty inches of rainfall per year. A large round, concrete in-ground tank from the old treatment facility gained new life as a stormwater storage tank when the city's engineering group redirected some of the existing storm drainage lines into it – effectively providing more than 3.1 million gallons per year. The stored stormwater is used for irrigation, flushing toilets, filling street-washing vehicles, and washing the fleet of white municipal vehicles. Equally innovative, the project team added a water-source heat pump to the design, taking advantage of thermal energy in the stored stormwater to augment the facility's winter heating and summer cooling requirements.

"A moving wall of light, now here, now there" [1]

In 2001, KEEN opened its eighth office in Canada, in Ottawa, Ontario, continuing to establish itself as a truly national firm. And, like Canada's famous Northern Lights, it was only a matter of time before KEEN Electrical would appear – a key piece of the strategy in becoming a multi-disciplined firm. In February 2001, the Electrical Group was formed in Edmonton to work with KEEN's energy contracting partner, VESTAR Inc.

Initially, KEEN Electrical was a group of one – Tony Grice, an electrical engineer who had recently left another firm after its purchase by a different company and was then invited to join KEEN. Grice's previous experience had included a number of green projects, serving as electrical engineer for the new Hinton Town Hall in collaboration with KEEN.

"Bear in mind, we were still predominately a mechanical engineering firm," recalled Kevin Hydes. "We agreed to add electrical only after discussion and consensus and good feedback from key people in the firm. But there were certainly those who didn't think we should ever go into electrical because we had never done it before. Tony was interested in becoming KEEN's first electrical guy because of the green agenda. If he hadn't joined us then, we probably still wouldn't have an electrical department – he was the bridge we needed."

And another twenty-five employees would soon follow. Among them was Met Ulker, with a long history of collaborating with KEEN on projects, who joined KEEN Engineering in late 2001.

> We've got expressways to information, and we've got side roads, and we've got back lanes. I can't say in all honesty that we've got the full road map figured out by any means. I think we're still building the knowledge network.
>
> – Kevin Hydes, KEEN

Based in Vancouver, Ulker is proud of his group's accomplishments over the last three years as part of KEEN. They include consulting on the Seattle Monorail Project (SMP), one of the largest public works in the city's history and due to open in 2009. KEEN engineers were retained to document the performance of mechanical and electrical requirements for all fixed facilities: nineteen stations and substations and the operations center. Another project, still underway, is the Arena Stage Expansion in southwest Washington D.C. Based on a design concept by Bing Thom Associates of Vancouver, the $100 million renovation will transform one of the top historic theaters in the United States into a sweeping, new performance space – with KEEN replacing nearly the entire electrical system.

Today, KEEN Electrical is well represented in the Calgary, Edmonton and Vancouver offices. Ulker is the group's business leader, while Grice serves as its technical leader. Both principals have worked to expand the role and focus of KEEN's electrical expertise into energy alternatives such as photovoltaics, wind generation, small hydro and even bio-mass applications.

In many ways, Tony Grice's own career mirrors the completely new breed of electrical engineer working today at KEEN:

> About fifeteen years ago while doing the Town Hall project in Banff, our energy simulations indicated that daylighting would be a key factor in the performance of the building. That was really mind-boggling to me, because I had spent the first ten years of my career trying to avoid daylighting. It was an intrusion to this controlled engineering space. It was interesting. As an electrical engineer why should I care about daylighting? My job was to put electrical lighting into a building, not worry about daylighting.
>
> So that project changed how I looked at things. Suddenly, I realized how a building was oriented on the site, how the daylight came into the building when it did and how incredibly important it was to the building. Now every time we design a building, we try to eliminate the electric lighting – that's doing our job.

Sustainable Concepts

By the Spring of 2001, partners Mitchell, McCarry, Hydes, and Anseeuw had achieved four of the launches in their five-part corporate strategy for KEEN: adding electrical services; establishing an energy performance partnership; opening new offices to create a national presence for the firm across Canada; and expanding into the U.S. That left one major piece: the establishment of a separate, green consulting team.

Hydes had demonstrated his knack for recognizing and seizing opportunities many times before. While attending a United States Green Building Council federal summit that April in Washington, D.C., he was introduced to Mario Seneviratne, originally from Sri Lanka and one of the first engineers to be accredited under USGBC's new LEED® program. A month later, Seneviratne contacted Hydes again to express his interest in setting up a new green group for KEEN:

> So here's a guy coming half way around the planet to join us, and later I asked, "Why do you want to work for KEEN?" We talked about the blue KEEN, the green KEEN. And he responded, "I'm going to help you create KEEN 3." What I admired about him was his ability to think outside of any parameters he was given.

Initially called "KEEN Green", the group was established to support sustainable design efforts in the Vancouver and Seattle offices. In less than six months, a team of half a dozen specialists were in place, recruited from top universities – and focused on new design techniques, advising KEEN's staff internally and consulting for outside clients.

"The arrival of KEEN Green was a big move for the firm because it created an internal team able to concentrate on sustainable issues," said Hydes. "And really, they were kind of an organizational change machine. Their goals involved more than selling services; they became a mechanism, if you like, for accelerating the internal transformation of KEEN."

In April 2002, the team was re-named the "Sustainable Building Service Group" and became a self-sufficient entity within KEEN. Now managed by Jennifer Sanguinetti, the group's mission had become more defined with experience:

> We've functioned primarily as a complement to the design side. We don't produce drawings. We don't do the nitty-gritty of design. We're there to support design teams through our modeling services and help green designs get better. There are so few people in North America doing what our team is doing.
>
> Being part of an engineering company – as opposed to an architectural or an entirely independent consulting firm – gives us a unique perspective with facilities staff, because we can talk knowledgeably about the controls and maintenance issues.

Known as "KEEN Concepts" since November 2004 – a name more fitting to its broader consulting role – the group today offers core

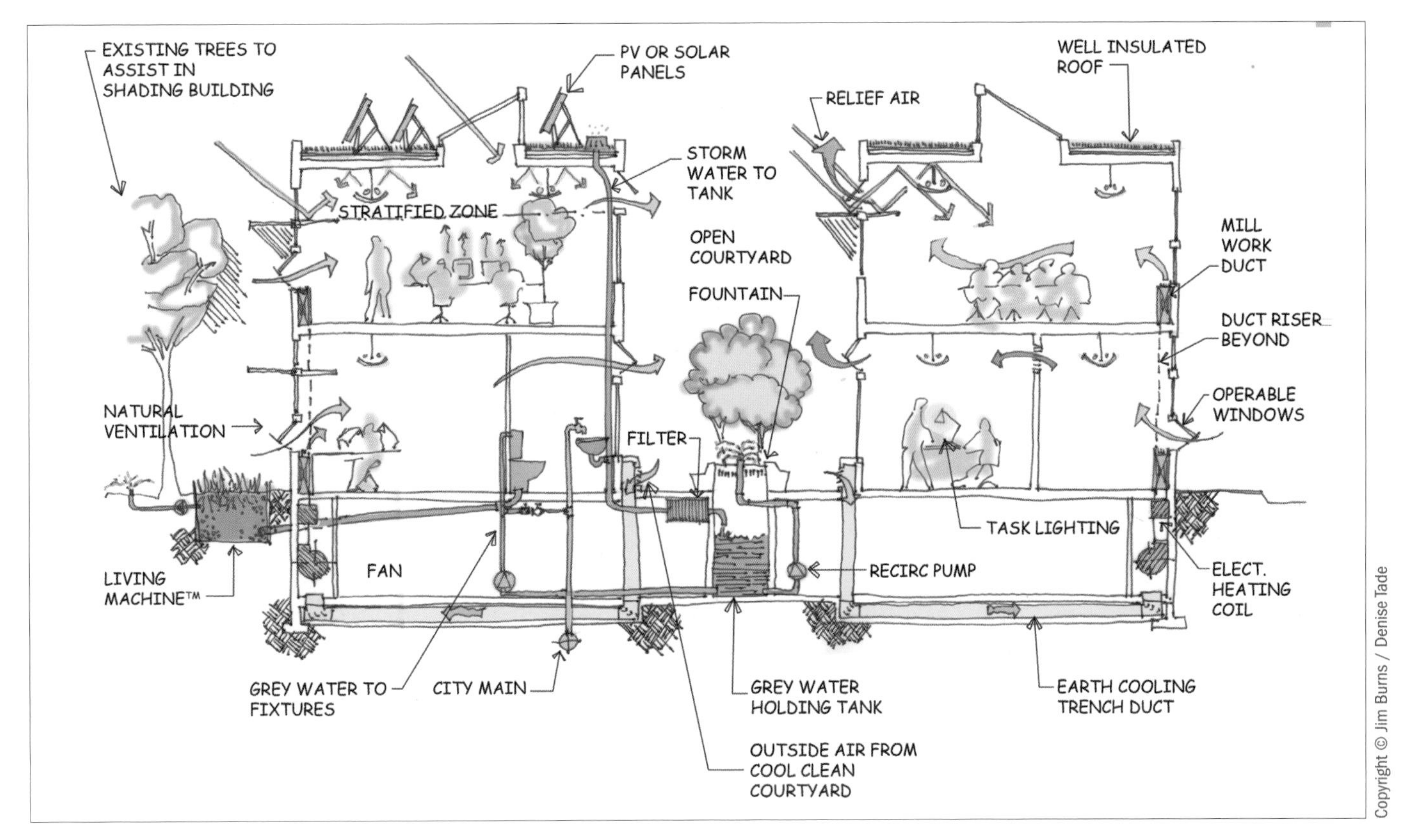

In coordination with BNIM Architects, KEEN's concept sketch described the full range of HVAC and water techniques proposed for the Packard Foundation's new 90,000-square-foot facility in Los Altos, California (2001).

competencies in water, energy and post-occupancy evaluations (POEs). Its services cover three main tracks:

- LEED® and sustainable design facilitation: to assist with schematic design and help manage projects through the LEED® process or simply green a building project
- Model building performance (energy, thermal comfort, natural ventilation, daylighting) and assist to architects with orientation and shading studies
- Existing building work: post-occupancy evaluations, LEED® for existing buildings, energy audits, water/waste audits, and advisory services to assist building owners in greening their operations

With a current staff of sixteen, KEEN Concepts has continued to make a name for itself across North America in helping to grow and improve sustainable building practices inside and outside of KEEN Engineering. That body of work includes a number of "firsts": consulting on one of the first LEED® -EB projects, the Moss Landing Marine Laboratories at Monterey Bay, California; LEED® facilitation for two future LEED® Platinum buildings - the Alberici Headquarters in St. Louis, and the Southface Eco Building demonstration project in Atlanta; and advisory services to the City of Whitehorse, Yukon, which invited Concepts to assist local architects in improving their green design practices.

New Tools, Wiser Choices

KEEN Concepts – and its earlier iterations – would also play a key role in developing several new sustainability tools with far-reaching implications for the green building industry.

In January 2001, the David and Lucile Packard Foundation announced plans to construct a new 90,000-square-foot headquarters in Los Altos, California. BNIM/Elements invited KEEN to submit mechanical services and ideas for its RFP on the project – which had an as-yet-unspecified goal of being green. Following a lengthy interview process, BNIM – and KEEN – were selected for the project team. Hydes was the principal-in-charge for the MEP services and Sanguinetti served as their project manager.

The Foundation identified USGBC's LEED® rating system as an appropriate measuring device for its sustainability goals – with the scoping document targeting LEED® Silver for the new office building. In preparation for the project's day-long charrette, the design team created a matrix showing how LEED® Silver was perhaps a conservative target for Packard. The design team recommended factoring in other criteria such as capital costs, operating costs and energy consumption as well, and that higher LEED® levels should be considered.

In response, Packard's representatives asked: "What exactly does this all really mean? What does it mean to build a market-rate building or to build a LEED® -certified building?" Their challenge to the design professionals – right in the midst of California's Energy Crisis, as energy prices continued to escalate overnight – was to provide a decision-making tool that clearly explains the aesthetic, economic, scheduling and environmental implications of the project.

Three months later, the project team had produced the Sustainability Report and Matrix, outlining schematic designs for six building scenarios: the four LEED® levels, plus conceptual building model characteristics for a "Living Building" and a "Restorative Building". KEEN's role was to develop energy models and life-cycle costs based on data from a cost consultant, and then produce the electrical and mechanical narrative on what each of these buildings might look like. For each scenario designed and defined by the project team, KEEN performed energy simulations and estimated construction, utility and maintenance costs; then KEEN forecast long-term costs using thirty year, sixty year and one hundred year cost models.

> Green buildings do not poison the air with fumes nor the soul with artificiality. Instead, they create delight when entered, serenity and health when occupied, and regret when departed. They grow organically in and from their place, integrating people within the rest of the natural world; do no harm to their occupants or to the earth; foster more diverse and abundant life than they borrow; take less than they give back.
>
> – Natural Capitalism: *Creating the Next Industrial Revolution*

By the time Packard's Sustainability Matrix was completed in late 2003, Sanguinetti recognized its importance not only for green buildings but in KEEN's larger, growing role of educating clients:

> It was a huge leap forward in terms of the information we could give to building owners. The matrix also moved us from being doers to being teachers – to educate CFOs, CEOs and other decision-makers about the impact of a mechanical system or architectural shading options and other choices on a building – and how a building reflects their company values.

Since becoming the first in the U.S. to formally adopt a city-wide sustainable building policy in 2000, the City of Seattle had explored

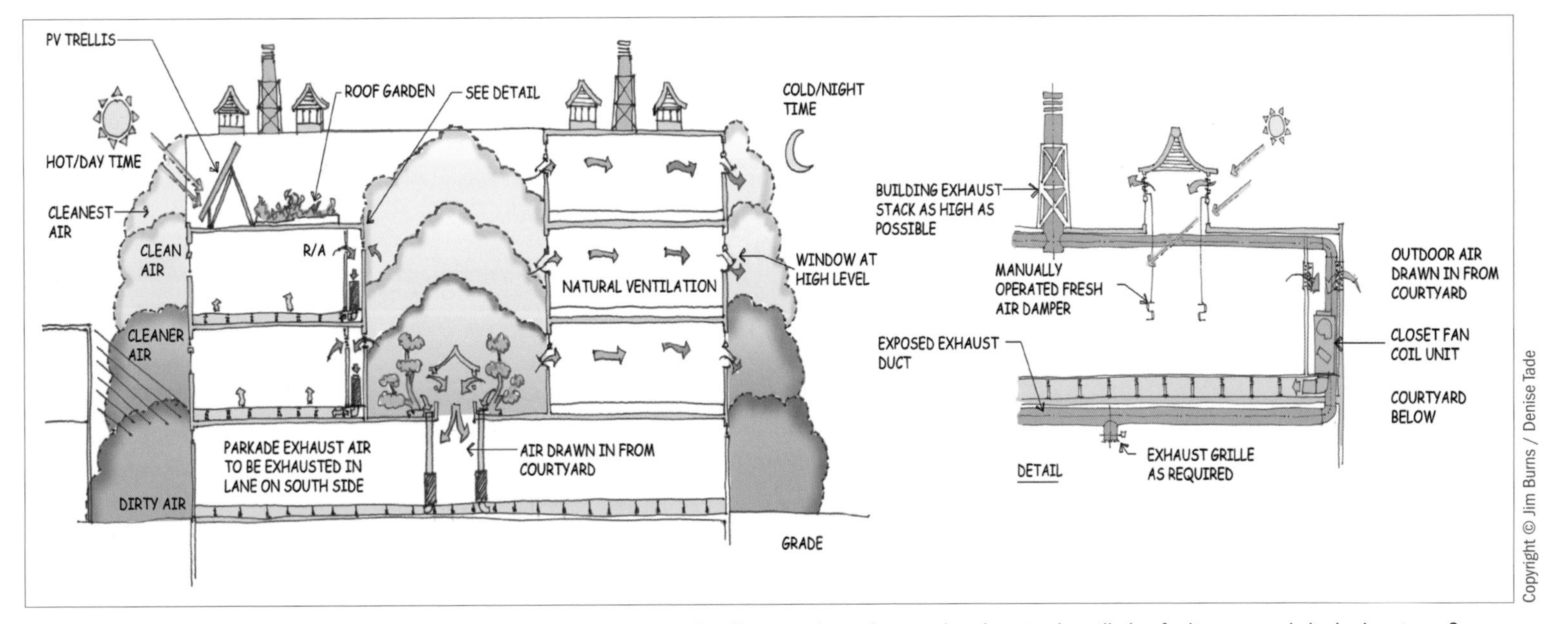

KEEN concept drawings for the Osher Center for Integrative Medicine were used to illustrate clean air strategies via natural ventilation for its proposed site in downtown San Francisco.

various resources to help building industry professionals understand and apply design strategies appropriate for the area's climate. The USGBC's LEED® Silver was mandated for any of its building types. The city wanted a web-based tool that would be accessible, user-friendly and content rich.

Again, BNIM Architects/Elements invited KEEN to participate in the RFP, and together they won the project commission in mid 2002. Named "Implement", the final web site represented the first time that KEEN Concepts had used its modeling capabilities as a significant part of a public education tool.

Noted Sanguinetti: "It was probably the first time that anyone had actually sat down and done that kind of full analysis of what you should do: for this climate, for this building type, and in that level of detail."

For this project, KEEN's task was to perform energy, orientation and natural ventilation modeling in conjunction with Elements on internal modeling, to calculate all cost implications for several different building types. The resulting on-line tool evaluates various sustainable design strategies, such as the impact of high or low thermal mass in the building on natural ventilation in terms of both air change effectiveness and energy and thermal comfort. Included are "Integrated Design Tools" for water and energy, "Case Studies", "Resources" and "Best Practice information", all sorted by project type. Implement also takes users through a five-step process informing owners, designers, engineers, contractors, developers and others on how to create a better building.

Concurrent with the Implement web site project, KEEN Concepts was invited to San Francisco to consult on a new building for the Osher Center for Integrative Medicine on the UC-SF campus. The client was familiar with KEEN's work on the Packard Matrix the previous year and wanted to "green" its proposed facility. In the charrette kicking

Moss Landing Marine Laboratories

Location:	Moss Landing, CA
Type:	University, Marine Laboratory
Size:	60,000 sq.ft.
Completion:	2004
Rating:	LEED®-EB Gold Certified
Architect:	Existing building, no architect involved

Sustainable Features:

- Conservation and restoration of native plant species
- Encouragement of natural pest management including birds of prey and foxes
- Discovery of the legless lizard, a burrowing species previously placed on the U.S. Endangered Species List; ongoing protection and species study
- Education outreach program – Friends of Moss Landing
- 99 percent reduction of potable water use for landscape irrigation
- 50 percent over waste stream diversion rate
- Low environmental impact cleaning policy
- Exceptional use of vision glazing, daylighting and exterior shading devices
- 5-year HVAC system upgrade strategy
- Reduction in energy consumption by 55 percent over ASHRAE 90.1 standards

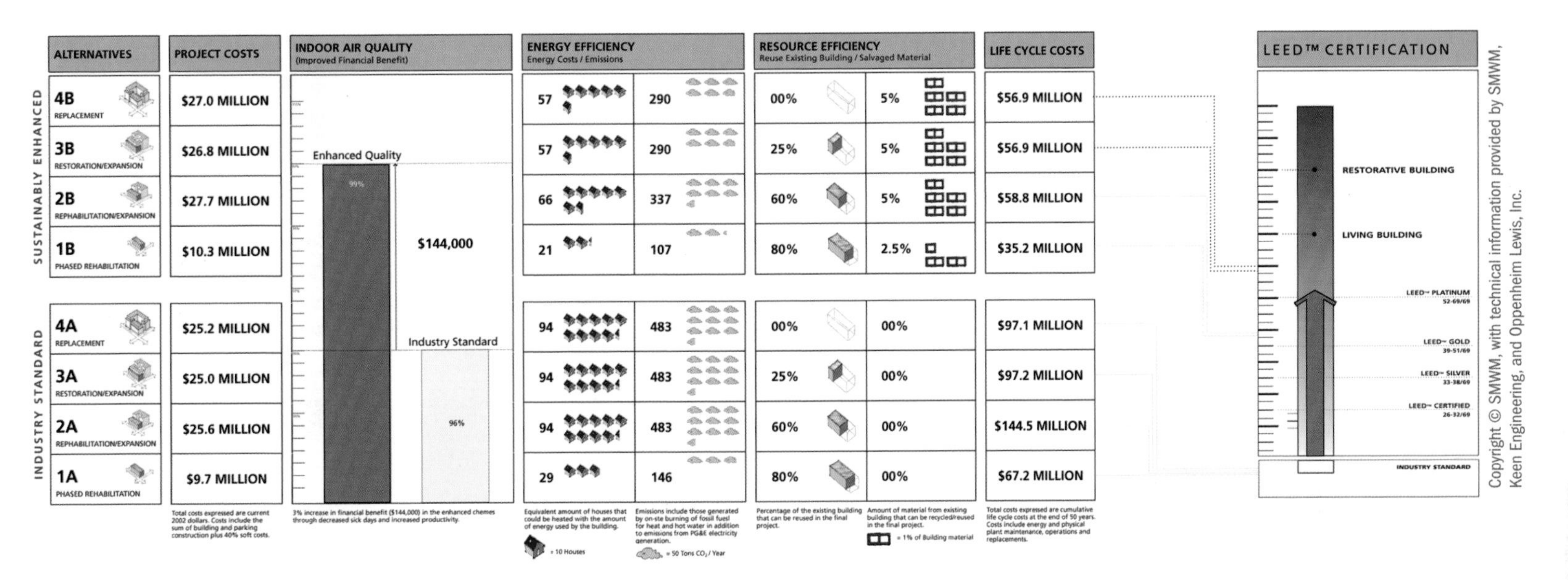

	ALTERNATIVES	PROJECT COSTS	INDOOR AIR QUALITY (Improved Financial Benefit)	ENERGY EFFICIENCY Energy Costs	ENERGY EFFICIENCY Emissions	RESOURCE EFFICIENCY Reuse Existing Building	RESOURCE EFFICIENCY Salvaged Material	LIFE CYCLE COSTS
SUSTAINABLY ENHANCED	4B REPLACEMENT	$27.0 MILLION	Enhanced Quality 99% $144,000	57	290	00%	5%	$56.9 MILLION
	3B RESTORATION/EXPANSION	$26.8 MILLION		57	290	25%	5%	$56.9 MILLION
	2B REPHABILITATION/EXPANSION	$27.7 MILLION		66	337	60%	5%	$58.8 MILLION
	1B PHASED REHABILITATION	$10.3 MILLION		21	107	80%	2.5%	$35.2 MILLION
INDUSTRY STANDARD	4A REPLACEMENT	$25.2 MILLION	Industry Standard 96%	94	483	00%	00%	$97.1 MILLION
	3A RESTORATION/EXPANSION	$25.0 MILLION		94	483	25%	00%	$97.2 MILLION
	2A REPHABILITATION/EXPANSION	$25.6 MILLION		94	483	60%	00%	$144.5 MILLION
	1A PHASED REHABILITATION	$9.7 MILLION		29	146	80%	00%	$67.2 MILLION
		Total costs expressed are current 2002 dollars. Costs include the sum of building and parking construction plus 40% soft costs.	3% increase in financial benefit ($144,000) in the enhanced chemes through decreased sick days and increased productivity.	Equivalent amount of houses that could be heated with the amount of energy used by the building. = 10 Houses	Emissions include those generated by on-ste burning of fossil fuesl for heat and hot water in addition to emissions from PG&E electricity generation. = 50 Tons CO_2 / Year	Percentage of the existing building that can be reused in the final project.	Amount of material from existing building that can be recycled/reused in the final project. = 1% of Building material	Total costs expressed are cumulative life cycle costs at the end of 50 years. Costs include energy and physical plant maintenance, operations and replacements.

SMWM's Osher Matrix examined the performance impacts and capital costs of four distinct scenarios – ranging from building renovation to complete new construction.

off the project, the client's top priorities were indoor air quality, energy and LEED® Platinum. In addition, they needed to address an existing building on the site of the new center.

As a result of its detailed study, KEEN Concepts presented a full decision-making matrix with SMWM to the Osher Center, proposing four separate schemes. These schemes included: refurbishing the existing building; constructing a new building essentially on top of and around the existing building; retaining the existing structural shell with a new building inside; and finally, demolishing the existing building and constructing a new one.

In helping to create the Osher Matrix, KEEN consultants needed to determine performance impacts and capital costs in all key areas for the four scenarios. That assessment meant examining the feasibility of energy performance in targeting LEED® Platinum, the impact of indoor air quality, quantifying pollution impacts, and working closely with the cost consultant throughout the project.

"This project really provided us with an incredible example of the impact of life-cycle costing," said Sanguinetti. "And it was yet another example of how working with the client early in the decision-making chain can exert a much greater influence on what systems and architectural decisions are ultimately made."

"You're only as good as your last shift."

Created in 1999 by the U.S. Green Building Council, LEED® has become the industry standard, an environmental report card, for rating new and existing commercial, institutional and high-rise residential green buildings. While attending a USGBC workshop in Seattle in early 2001, Hydes was so impressed with the just-launched LEED® system that he returned immediately to KEEN's office and gave a brief presentation that afternoon on what he had just learned about the new measurement tool. For several months, he continued to send regular e-mails around the office, encouraging staff to take the new LEED® exam and become certified. But never a response.

So one spring day, Hydes drove to the exam center in Seattle and took the exam, then e-mailed the entire staff upon his return to Vancouver: "Folks, we have now have one LEED® -Accredited Professional.' About twenty seconds later, I got an email back from Rosie Hyde, one of our newest employees, 'Actually, Kevin, we've got two.' It turned that out I passed the exam two hours before she did – otherwise, I would be wearing number 2 and she'd wear number 1." Those numbers refer to KEEN's LEED® hockey jersey, created as an incentive and a sign of accomplishment for any employee who passed the exam – complete with his or her name and number signifying the order in which the individual became LEED® -Accredited within the firm.

"We took an early position on LEED®," noted Hydes. "I know a lot of other firms did too, but we didn't just join the U.S. Green Building Council. We took it a step further by stating that LEED® would be in the center of our organization, that it's not just the responsibility of a partner or one individual member of the firm to bring back this information to the rest of us. So we kind of placed a flag in the middle of the firm and said that's what KEEN stands for now."

Those bright green hockey jerseys have also become a symbol of KEEN's corporate commitment to LEED® . At recent GreenBuild Expos, KEEN's booth featured some photos of award-winning buildings – but most prominently displayed were the jerseys of LEED® -Accredited employees. At every KEEN management conference, too, LEED® is the first item of business. Discussion among the partners, principals and associates in attendance focuses on the sustainable agenda, and how quickly the firm has progressed towards a common goal: "To be the leading sustainable engineering firm in North America. It's a bold statement to make, I know that," said Hydes, "but I think we've made some significant progress at it. And clearly LEED® has been a big part of trying to get there."

By early 2005, a total of 163 KEEN employees across the company had earned their LEED® jerseys. And now a growing number are gaining LEED® Version 2.0 certification – a tougher, more progressive

Courtesy of Keen Engineering

KEEN Principals Kevin Hydes and Paul Anseeuw hosted one of the first-ever USGBC-LEED® workshops in Seattle in 1999.

exam designed to reach still deeper into the details and data of the rating system. As a result, KEENers are showing up in the new, white V2 (for Version 2) hockey jerseys. So the challenge continues.

Said Hydes, "By now, everybody knows what our move's going to be, right? We're off again, it's another competition. And it's important we keep on doing this because, as they say in hockey, you're only as good as your last shift."

More Gold, in Monterey

Even in February at Moss Landing Marine Laboratories (MLML), it is possible to see and hear birds everywhere: pelagic seabirds skimming over Monterey Bay, hawks soaring along the rugged California coastline. It is a dynamic setting for this educational and scientific institution situated on a twenty-one acre hillside in the fishing port of Moss Landing, halfway between Santa Cruz and the Monterey Peninsula. A part of the California State University system, this 60,000-square-foot facility contains scientific laboratories, classrooms, a research aquarium, and a research museum and library. Students and faculty are engaged in a variety of Master's and Ph.D. marine research projects, both locally and globally.

Through Hydes' growing involvement with the USGBC, he met MLML's director, Kenneth Cole, at a LEED® -EB (Existing Buildings) kickoff in Washington. The USGBC had developed this new rating system so LEED® buildings could become re-certified and more broadly address the entire existing building stock out in the market. KEEN and the lab agreed to partner on a LEED® -EB pilot beginning in June 2002.

MLML wanted to make the building a showcase and a benchmark for new and existing buildings in the CSU system. The project had three primary objectives. First, the client was confident of their site restoration efforts, and much native wildlife had returned to the area after construction. But the client was less confident of the building's environmental performance and sought third-party verification on its design and operations. Second, the client saw LEED® -EB as a viable way to improve the lab's indoor environment. And third, MLML suspected there may have been some late programming changes to the building's layout without corresponding changes to its mechanical design. Dan Roberts, a sustainability consultant in KEEN's Vancouver office, became the project manager on the pilot, working closely with Barry Giles, MLML's Building Services Engineer.

LEED® -EB's mandatory prerequisites include testing all building systems to ensure they meet the original design intent. KEEN's initial task was to perform a facility assessment and report on any operational issues, providing a baseline on which to prioritize near- and long-term system repairs and upgrades. Number one on the list was temperature and humidity control issues in the Laboratory's research library, caused by too much outdoor air entering into what should have been an isolated space. To provide tighter climate controls for the library, Roberts proposed re-working the ductwork and diverting two of the three air handlers to other systems in the building.

Noise issues were a huge problem as well. KEEN's facility assessment revealed extremely high velocities of air moving through building ductwork and out of diffusers – creating whistling, rattling, rumbling sounds – in what were intended to be quiet office spaces. The design and layout of the ductwork for the building appeared correct, but Roberts discovered it was not consistent with the actual installation: huge ducts, large airflows shooting quickly through a small transition and diffuser. The installation problem was easy to fix with a few simple changes to the ductwork and return air construction.

The Laboratory's original commitment to site restoration and green building strategies included waste stream diversion, daylighting and exterior shading devices and later, HVAC system upgrades and improved indoor environmental conditions. Through these commitments MLML has served as an exemplary building for the LEED® -EB Pilot program, ultimately helping it to achieve LEED® Gold.

Beyond Knowledge

With the firm's rapid growth and fast pace on most projects, it has become a bigger challenge for KEEN engineers to share information,

Courtesy of Keen Engineering

KEEN's LEED®-Accredited Professional management team proudly wore their "KEEN/LEED®" jerseys at their Spring Management Conference, Semiahmoo, Washington.

Canmore Civic Centre

Location: Canmore, Alberta

Type: Municipal Government Office and Town Hall Complex and Museum

Size: 2,322 m^2

Completion: July 2004

Rating: LEED® Silver Certified

Architect: Marshall Tittemore Architects

Awards: 2005 Award of Merit, Consulting Engineers of Alberta
2004 Alberta's Construction Project of the Year, *Alberta Construction Magazine*

Sustainable Features:

- High efficiency (94 percent) condensing boilers
- High temperature drop heating system to reduce flow rates, decrease pipe size and lower distribution energy use
- The use of exhaust/supply air heat recovery
- Lighting zone controls and occupancy sensors
- A natural ventilation strategy which incorporates operable windows in all occupied spaces and narrow floor plates connect to the central atrium. Both cross ventilation and stack effect contribute to this strategy
- Dual-flush toilets, low-flow fixtures

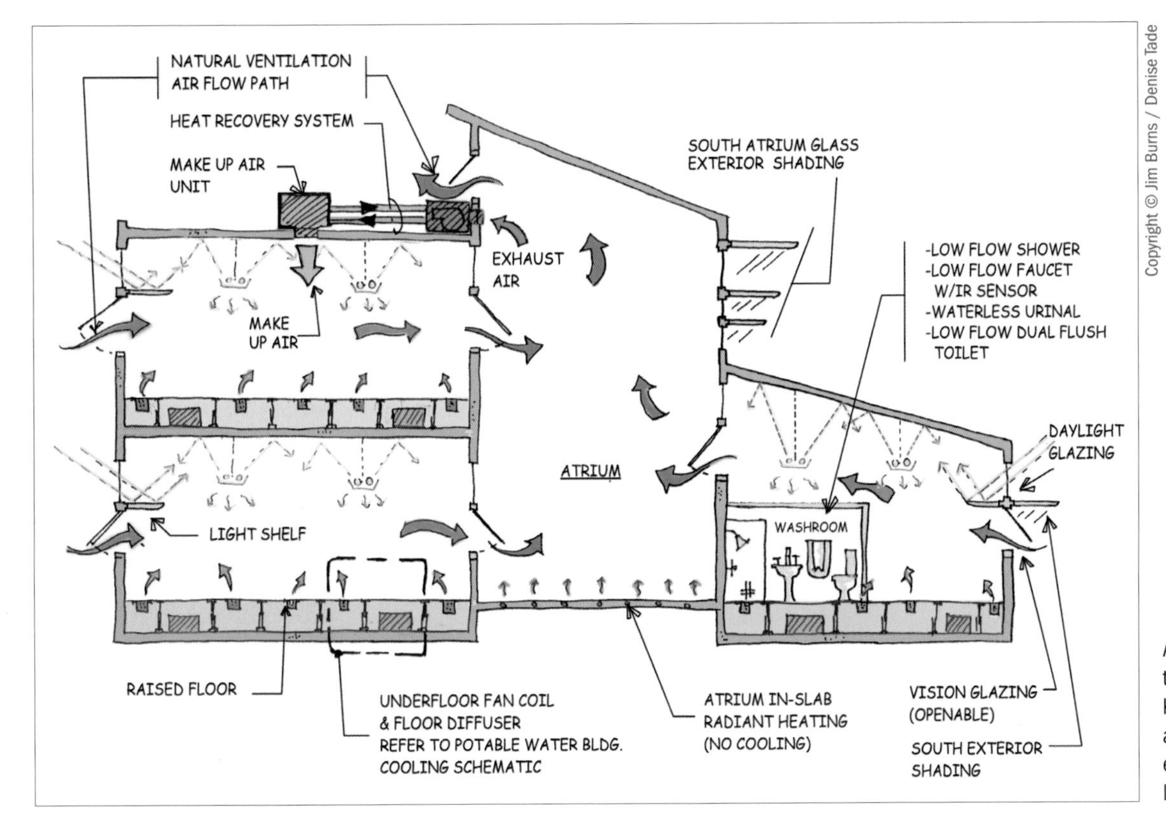

As part of the integrated design process for the new Canmore (Alberta) Civic Centre, KEEN's cutaway concept sketch proposed a wide range of green mechanical and electrical schemes. The facility was certified LEED® Silver, the first in Alberta.

to transfer knowledge company-wide, particularly in the rapidly evolving world of green design.

"In the old days, our technical library resided in the office of the senior partner of the firm," said Hydes. "That was the traditional model. If you needed to learn about something you'd knock on the door and ask, 'How do I design an XYZ system for a commercial building?' They would pick out a book, open it to a chapter and say 'Read this and come back to me if you've got any questions'."

More recently, KEEN has relied on its knowledge leaders scattered around the firm to provide answers and guidance on specific issues via e-mail or phone. For Hydes, that quick, on-the-go approach has worked just fine:

> That chain of information often gets me where I need to quickly. But the approach is also pretty limited because it assumes these five or six leadership cells have access to all the information 250 employees are doing – which is no longer true, given the amount of work going on. So it's not going through the hands of three or four or five or even a dozen people anymore.

In the last five years, the development of a knowledge network, a virtual data repository, has created a whole new way of learning, changing the way staff seek out and access information. As a multi-office firm, it is more important than ever to ensure that knowledge is shared between the different groups over this network, allowing designers, researchers, and consultants to upload or download the latest thinking on green techniques and issues. For Hydes, there is still no substitute for sharing knowledge one-on-one. Whenever KEEN's knowledge leaders visit another office, for example, they are expected to give a presentation on their current work and provide staff with the latest thinking on daylighting, thermal mass, and any of a dozen other topics. "To be honest, I think we're often so data overloaded that people may be more likely to recall that 'brown bag lunch', even if they don't use that knowledge for a couple of years. If they do get a new project requiring the application, they remember that event and the nuances that were given. Those presentations don't just dump data: it's 'Here's the data, and this is what it means.'"

> Our role as the engineer is not just to work with the architect or do the engineering – it's working with the authorities and pushing them and even stretching ourselves. In the end, we're changing the code. As I see it now, that's part of our job in these buildings.
>
> – Jim Sawers, KEEN

KEEN also recently set up a peer review process to address regulations in jurisdictions where a third party within an engineering firm is required to review all designs. KEEN's senior management saw this process as not only useful in promoting quality and innovation, but also a powerful way to get its own designers to study the work of their peers.

Alberta Breakthrough

The Canmore Civic Centre is a multi-purpose facility for the Town of Canmore, Alberta – a growing recreational and adventure center located in foothills of the Rocky Mountains, halfway between Calgary and Banff. In replacing its old town hall, Canmore wanted to develop a showpiece to promote sustainability in this pristine mountain environment.

KEEN served as both the mechanical and the electrical consultant, working closely with Marshall Tittemore Architects and the design team to deliver a fully integrated, sustainable facility. Together, they shared ideas and design concepts with the client over a two-and-a-half year period. Additional funding then came from Canada's Green Municipal Enabling Fund.

When the project commenced in June 2002, all team members and client representatives participated in an integrated design process, unanimously agreeing on a goal of LEED® Silver. At project's end, that goal did not affect either the pre-determined budget or completion date.

KEEN's Jim Sawers, of the Calgary office, served as lead engineer and principal-in-charge, with Matthew Peachman as mechanical design engineer. They produced a cutaway-style cartoon to help illustrate which green technologies should be applied in the building, and later produced a second cartoon of the project exactly as it was built. Observed Sawers:

> The two are remarkably similar, and I think that speaks to the fact that the engineering had a fairly profound effect on the architecture of the building. So the big move was the architect buying into the initial cartoon we drew up and saying, "Yeah, that's something we could make work." He was very, very open and picked up on these concepts immediately and worked with us to build it through to fruition.

In addition to considering orientation, massing, construction and sustainable transportation measures, a variety of green electrical and mechanical strategies were employed. Most notable is the use of the town's potable water system for building cooling, which even in August is only about 39°F. The water is taken from the town water main on one street, run through the building's cooling system and returned to a second town water main. This strategy avoids mechanical refrigeration, the associated energy use and the use of CFCs or HCFCs in the building's cooling system.

"No one had really done this before, so it became an issue of managing the risk in terms of the safety of the water supply," said

BC Cancer Research Centre

Location: Vancouver, BC

Type: Research Center, Laboratory

Size: 220,000 sq.ft

Completion: January 2005

Rating: LEED® Gold Certified

Awards: 2005 Consulting Engineer of B.C. Award of Merit

Architect: Henriquez Partner/IBI Group Architects in Joint Venture

Sustainable Features:

- Heat recovery chiller
- Energy efficient electric chillers
- Chilled slab flooring in the offices
- Low-VOC paints and finishes
- Dual-flush toilets, waterless urinal and low-flow faucets
- 42 percent energy savings (ASHRAE 90.1)
- 43 percent saving in potable water use
- DDC system controls motorized operable windows at the bottom and top of the spiral staircase; Windows are opened to relieve heat in the stairwell, eliminating need for mechanical cooling
- 26 percent recycled/reused materials
- variable volume fume hoods

Sawers. "We spent a lot of time working with the code authorities and bringing in specialists to add technologies for treatment and monitoring. In the end, we were able to make everyone comfortable with the system, so that someone at some subsequent time could drink the water."

Additionally, a raised floor displacement mechanical ventilation system was used throughout the Canmore Civic Centre, enhancing indoor air quality at lower operating costs through improved ventilation effectiveness in the occupied zone. Extensive use of low-flow plumbing fixtures reduced building water consumption by approximately 55 percent compared to a standard building. High efficiency condensing boilers, exhaust/supply air heat recovery, occupancy sensors, daylighting and natural ventilation all contributed to maximizing energy efficiency. Overall, building energy performance has been confirmed to be 40.4 percent better than the baseline building represented by the Canadian Model National Energy Code for Buildings. Recognized as the first LEED® Silver building in Alberta, the completed two-story, 25,665-square-foot Civic Centre now serves as the town hall, the municipal government office, council chambers, and the new Canmore Museum and GeoScience Centre.

When Form Meets Function

The new 230,000-square-foot BC Cancer Research Centre is one of the largest freestanding cancer research facilities in Canada. Its fourteen stories rise over the south slope of False Creek in Vancouver and make an unmistakable impression on the city's downtown skyline. Its art moderne, slatted, circular windows – sixty-eight in all – are a reference to petri dishes. The design by Henriquez Partners/IBI Group Architects Joint Venture also features an external spiral double helix staircase connecting the office tower to the sustainable lab building, where the mechanical equipment is sandwiched between each lab floor. Every window is fitted in colored glass strips, emulating chromosome 8. And the top-floor meeting room has an amoeba-shaped roof. "The Petri dishes, the DNA staircase, the chromosome 8 glass, the amoeba roof – all are architectural iconography, and they are what will stake this building's claim in Vancouver's history," reported The Vancouver Sun.[2]

Representing the most successful health-care campaign in British Columbia's history, the millennium campaign raised nearly $100 million towards the construction of this project. The BC Cancer Research Centre is expected to attract and retain the world's top scientists and facilitate close working relationships between scientists and clinicians.

Henriquez Partners brought KEEN's Vancouver Office onto the project in October 2002. Together with the building owner, the BC Cancer Foundation, they agreed the design goal for this project would be to achieve a LEED® Silver rating, the first of its kind for a health care facility in Canada. It also is designed to be adaptable for future technology and innovations.

KEEN's design contributions included energy-saving measures such as use of a heat recovery chiller and chilled slab flooring in the offices. Low-VOC paints and finishes, water efficiency measures, materials recycling and procurement, and construction waste management contributed to the building's sustainability. Innovation credits included the use of interstitial spaces between each lab floor for the electrical and mechanical systems to provide long-term flexibility in the building and minimal disruption to adjacent laboratories during future renovations.

Constructed in two phases, the facility houses four hundred researchers in eight research departments, conducting research into all of the two hundred known types of cancer. It will ultimately house six hundred researchers. One of the specialty laboratories includes the Michael Smith Genome Sciences Centre, the only genome laboratory in Canada dedicated to cancer research. The B4 basement level houses one of the largest vivariums in Western Canada, in 10 percent of the overall project area.

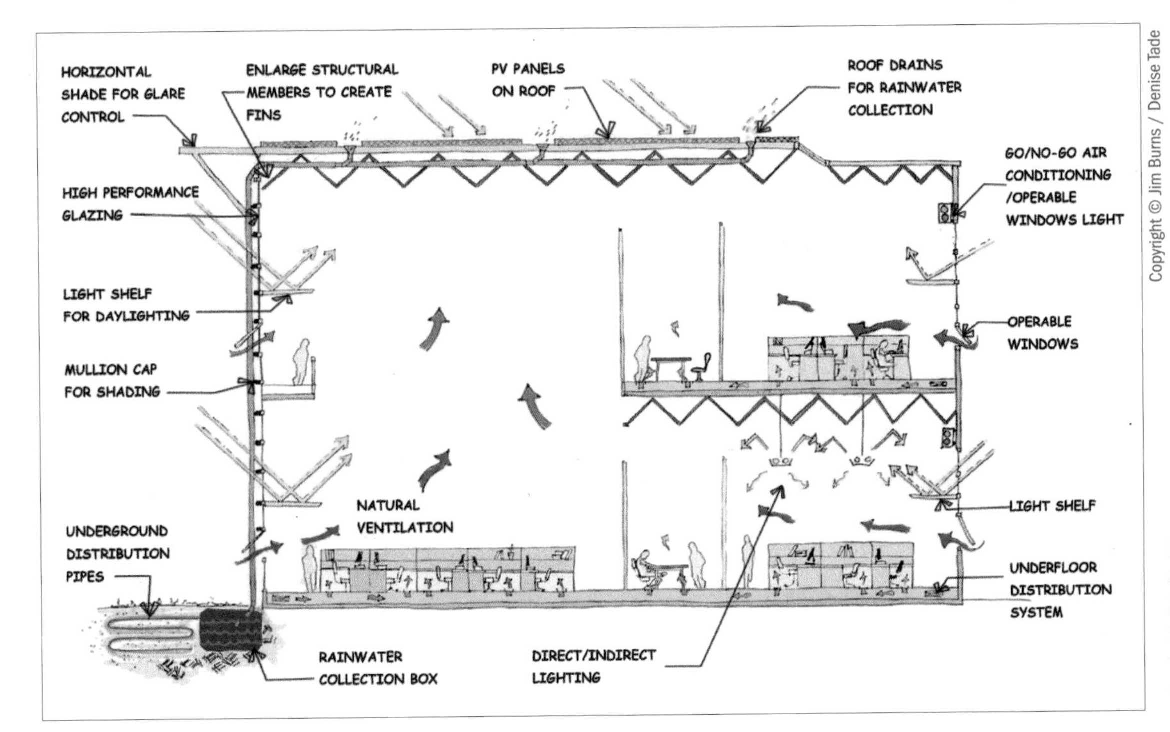

Copyright © Jim Burns / Denise Tade

KEEN played an instrumental role as sustainability consultant on the new headquarters for Alberici Constructors of St. Louis. The concept sketch highlighted the many design features that led the facility to LEED® Platinum.

Influencing the Industry

Being KEEN green has also meant a growing ability – or even a necessity – to influence an issue, to become a catalyst in transforming the market itself. It did not start out that way. Initially, KEEN simply wanted to influence the people who might purchase its green services, realizing at the same time it was necessary to influence the firm's own designs. But even as far back as the early green projects, McCarry, Hydes and others started to collaborate actively with several ASHRAE scientific committees on draft standards and practices for underfloor air distribution systems and natural ventilation.

Still, KEEN was getting concerned about the "black magic effect" of some of its early work. Hydes stated:

> We couldn't quantify some of the interactions going on between a system and the building and the people. Then we started to get our own direct feedback and experience, which drew us into these committees. Their early drafts supported what we were seeing on our projects and, in turn, allowed us to detail the next wave of projects. So it became a two-way exchange of giving and receiving information with the committees, leading to standards that are now issued for the whole industry.

More recently, KEEN has influenced codes in several Canadian jurisdictions, and even a draft national building code of Canada that included no mention of natural ventilation. As a result, the firm wrote to officials, requesting the necessary changes and additions through the technical appeals process – and changes were made.

Professional ethics have become a consideration for the firm, as well, in the relatively new, still-evolving green building industry. In response, KEEN was instrumental in convincing the thirty-thousand member Professional Engineering Society of British Columbia to include LEED® in its by-laws.

"There's a standard engineering creed to protect the public, protect the environment, protect the client and protect the profession," explained Hydes. "So British Columbia is the only jurisdiction anywhere in North America at this point that says protect the environment, with some guidance as to what that means, look at the LEED® standards, etc. That's fairly significant."

"Influencing change is a responsibility the firm and members of the firm take it seriously," he added. "To me it's our obligation to get involved with these committees, councils and societies as well as practitioners and educators – to really start to engage people who are living and working in buildings."

Showcasing Green: St. Louis

In December 2004, 180 employees of Alberici Constructors, Inc., eagerly moved into their new office space outside of St. Louis, Missouri. Once an abandoned, derelict industrial site, the park-like headquarters now features the showcase Class "A" office building, a covered parking garage, outdoor dining area, all within a park-like setting of ponds, wetlands, native grass prairie and walking paths.

"The building is an employee's dream," wrote the *St. Louis Construction News & Review* upon its opening. "From every cubicle, office, training classroom, and seat in the full-service cafeteria, there's a view of the outdoors. Operable windows at occupant and clerestory heights admit fresh air and natural daylight. Employees have individual control over the temperature and airflow within their workspaces."This new building could bring Alberici international fame – not just recognition, but fame: the building stands a very good chance of being only the tenth LEED® Platinum- certified building in the world." [3]

As part of the Alberici Corporation, an international firm headquartered in St. Louis, Alberici Constructors is the third largest contractor in Missouri – and they needed room to expand. In November 2002, they had selected St. Louis-based architects Mackey Mitchell Associates to create their new corporate offices. Given the company's international work, senior managers at Alberici were well aware of the push for green projects in Europe and of the economic benefits of green design. Motivated by some of its own clients' effort to achieve LEED® certification in North American, Alberici decided the new building would be LEED® Platinum. So KEEN Engineering was a natural addition to the project team in completing design development and act as sustainable building consultants.

> We want to understand what it takes to do green buildings, to build one, to live in it, and to be better prepared to work with clients who also want to go green. We don't think this is just a fad.
>
> – Bob McCoole, President, Alberici Corporation

Jennifer Sanguinetti, of KEEN's Concepts Team, led the eco-charrette to kick off the project, joined by Tony Grice of KEEN's Edmonton office, as project consultant. Through the planning process, the Alberici management team defined five core goals for the project that, in addition to LEED® Platinum, included a goal that the project cost no more than a conventional building of the same size and style. "That was a real challenge," said Grice, "but KEEN's experience as an engineering firm is that it's not an unrealistic goal. It's very interesting because there was a vision of LEED® Platinum, but it was the reality of the balance sheet as well. In finding that balance, we relied on their experience as contractors – because if we proposed concepts we feel are viable for the project but that weren't going to work financially, they'd let us know right away."

With the help of the architects, Alberici found the perfect location at a major intersection, Interstate 170 and Page Avenue in the St. Louis suburb of Overland. They also found what they considered to be the perfect building on a fourteen-acre brownfield site: a fifty year old, three-story brick office building next to the 300,000-square-foot, all-steel warehouse of a former metals manufacturing company.

"They were making a statement with this project, by taking this old, gorgeous but filthy dirty, falling apart, steel manufacturing plant and turning it into this office building," said Grice. "That structure had so much inherent beauty in it – but you'd have to be a bit of an architect or artist to stand in the middle of this decrepit building to understand that. Fortunately, the architects saw that vision too. So they did some really great renderings of what the building could look like."

The architect's initial design utilized the entire floor plate and structure of the original building. KEEN's energy modeling and sunshading study for the project, however, revealed that the existing building orientation posed a huge challenge in attaining the LEED® Platinum rating. Grice suggested moving a planned wall of office windows on the south to the northern side. In considering the building's southwest orientation, the team also agreed on a sawtooth pattern to give the building a more favorable southern exposure and allow for controlled daylighting. Atriums and operable clerestory windows were added to further reduce the need for artificial lighting.

"Turning the building was the natural solution," Grice said. "It's interesting that we, as the engineers, actually changed the whole concept of the building, but in a very collaborative way...and that's all part of the integrated design process, which worked perfectly."

Upon the start of construction, Alberici crews disassembled the brick office building and were able to re-use, donate or recycle 95 percent of the building materials - brick, limestone, glass, aluminum, copper, stainless steel, concrete and roofing.

To create the new office space and parking garage, the old warehouse/plant was sliced lengthwise into three sections – removing a 70-foot-by-485-foot section of the roof in the middle, effectively creating two parallel buildings. The southern building, with its ninety-foot clear span, was converted into a 110,000-square-foot, mezzanine-style office space with two levels and three atrium areas. In the northern building, two of the seventy foot bays were transformed into a covered, two-tiered parking garage. The space between the two structures then became a breezeway and courtyard, with a portion of the original roof used as an overhead canopy.

The new Alberici Headquarters showcases many other features of green design and engineering. The HVAC system employs both natural ventilation and an underfloor air distribution system. A large trench once used for servicing trains now contains some of the air ducts and they benefit from the passive, below-grade warming or cooling. Water-saving measures include low-flow faucets, waterless urinals, dual-flush toilets, and a rooftop cachement system, which captures rainwater from the garage roof and stores it in a 31,000-gallon cistern – saving Alberici approximately 500,000 gallons of water per year.

To date, about 20 percent of Alberici's total energy requirements are met with renewable sources including a massive 65-kilowatt, three-blade wind turbine, solar panels that supply 90 percent of the building's hot water needs, and a ground-source heat-pump system connected to a grid of one hundred geothermal wells.

Demonstrating Green: Atlanta

For more than twenty-five years, Atlanta's Southface Energy Institute has been educating architects, homebuilders, other residential construction professionals and homeowners on sustainable building practices. Every year, visitors come to the Southface Energy and Environmental Resource Center – a model home in downtown Atlanta built in 1996 – to see more than one-hundred "real-world" solutions for environmentally-friendly living. With interest continuing to grow, the Institute decided to apply its expertise to the commercial building arena through creation of a regionally appropriate, high-performance office that would highlight cost-effective ideas for saving energy, water and other natural resources.

In 2002, this non-profit institute developed its concept for a new, state-of-the-art demonstration building known as the Southface Eco Office. The building was envisioned as a model of environmentally-responsible commercial construction and development – a relatively low-cost "green" commercial building that would foster a healthy and productive workplace.

Through a design competition, Southface selected Lord, Aeck & Sargent, Inc. of Atlanta as its architectural partner for a "design that is efficient and elegant, 'green' and functional". Southface also assembled a "construction consortium" for the project. And with the new office underway in June 2003, KEEN Engineering was retained to perform energy modeling and provide full mechanical, electrical, plumbing and fire protection services.

From the outset, the Georgia climate posed a real challenge for KEEN because of high dehumidification loads during the cooling season. Pasha Korber, of KEEN's San Francisco office, served as MEP project manager for the Southface Eco Office:

> We looked at a number of systems proven in the European building markets but not yet adopted and tested in the U.S. That meant considering "some controversial" applications not typically found in the southeast's extremely moist, humid climate: applications like chilled radiant ceilings; an atrium space with a chilled slab/ thermal mass on the first floor; fan coil units for ventilation; and an underfloor air distribution system.

The design team also chose to decouple the ventilation, resulting in a system that separates the ventilation load from the space conditioning loads and significantly lowers HVAC energy consumption. The excess moisture will be absorbed by the desiccant dehumidifier, and the desiccant material will be recharged using solar energy and waste heat to dry out the moisture.

KEEN's recommendations to the project architect include external shading, favorable building orientation and a high-performance

Rendering courtesy of Lord, Aeck & Sargent

Southface Eco Office

Location: Atlanta, Georgia

Type: Office (Commercial office/Demonstration office)

Size: 7,000 sq. ft.

Completion: 2005

Rating: LEED® Registered

Architect: Lord, Aeck & Sargent Architecture

Sustainable Features:

- Solar heated domestic hot water
- Extensive daylighting, external shading, building orientation
- Underfloor air system
- Green roof
- No potable water for waste conveyance or landscape irrigation
- Low-flow plumbing fixtures, including dual-flush toilets and waterless urinals
- Stormwater system for flushing toilets, composting
- Photovoltaics
- Dessicant dehumidification
- Dedicated VAV outside air system with CO_2 sensor control
- Airside heat recovery

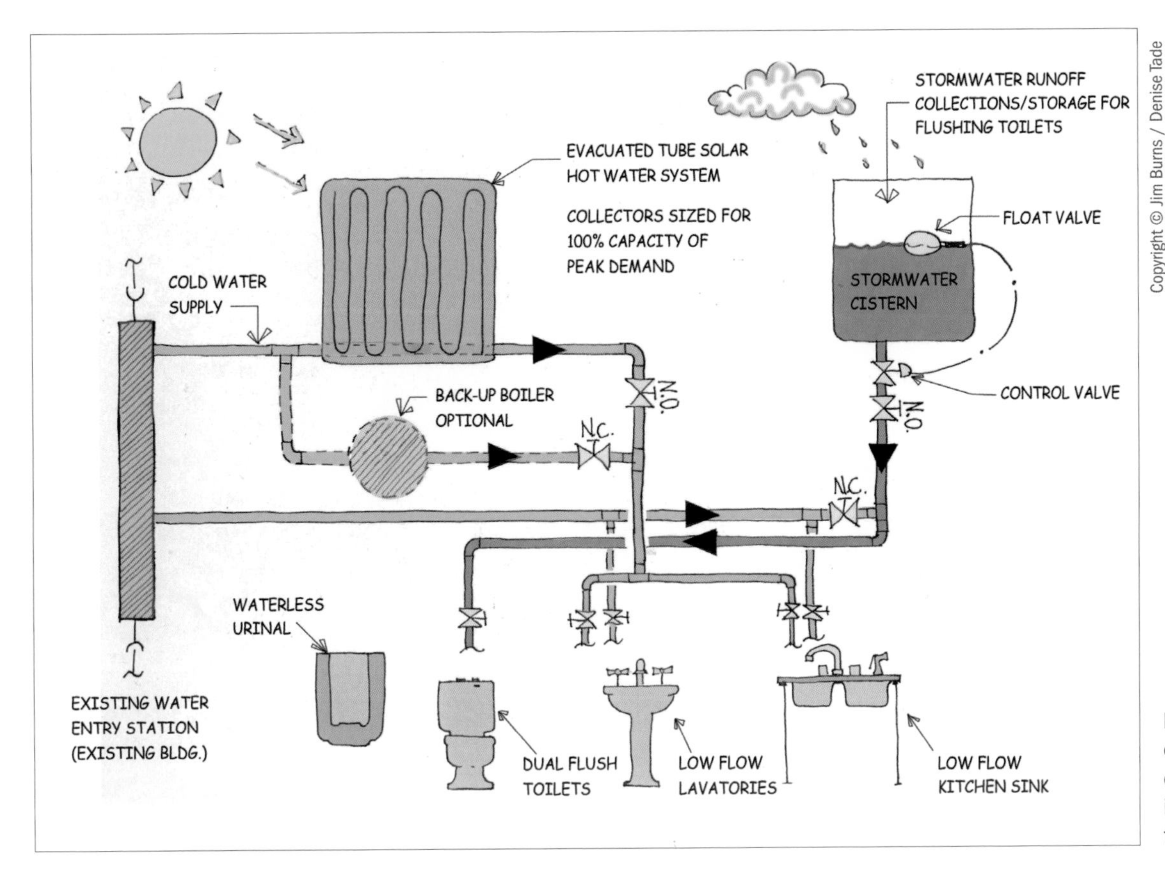

For the Southface Eco Office, a green demonstration office near Atlanta, KEEN developed an extensive water scheme to include stormwater collection, low-flow fixtures and a solar hot water system.

building envelope to minimize direct solar gain in the building and to significantly reduce cooling equipment capacity and system energy consumption. In addition, the office will generate a portion of its own electrical power on site with a micro-turbine cogeneration system. The design team understood that the final building must effectively help "teach" sustainable, yet attainable, design strategies for the southeastern U.S. Upon its expected completion in November 2005, the Southface Eco Office will be one of the first LEED® Platinum buildings in the United States.

Evaluating Green Buildings: Ipso Facto

"The building learns from its occupants, and they learn from it."[4]

KEEN Engineering could not agree more, believing that post occupant evaluations (POE) are invaluable tools for improved building performance. Under the guidance of Rosie Hyde, the firm has incorporated the Center for the Built Environment's occupant satisfaction surveys into its own POEs to gauge the effectiveness of building services and design features on several of its completed green buildings – and to prioritize the steps necessary to further improve satisfaction and workplace productivity.

And KEEN is also one of the very few consulting engineers in North America that actually asks clients and building users what they think of their buildings. The firm's evaluations include measuring energy and water usage as well. "It became clear that we needed to know

Rendering courtesy of Lord, Aeck & Sargent

Atlanta's Southface Eco office building has been designed to showcase innovative use of materials and state-of-the-art energy features – while effectively "teaching" sustainable, yet attainable, design strategies for the southeastern U.S.

what our performance levels were," said Hyde. "In learning from our own buildings, we needed a standard for designing them and a standard for operating them – did our green designs lead to truly green building operations? – and then feed what we learned back into our next designs."

Much of Hyde's research also involves developing a better protocol for POEs that will go into the public domain and seeking additional grants to fund future evaluations. Her focus too, is on collaborating with other building consultants and academics to gain their perspectives as part of building the firm's knowledge base. For KEEN, each new POE seems to further validate its green design approach and practices of the last several years: that occupants like daylighting and natural ventilation, and they like having personal control over their work spaces with elements like operable windows. There is also room for improvement in the areas of thermal comfort and acoustics. The POEs completed so far have also reinforced key aspects of the design process itself: that the building owner fully understands the design; that sufficient modeling is conducted; and that all users know how to operate it.

"To go back and look at whether a building is actually working or not and document it for the public – that's the kind of thing we're really learning, that's what we're so proud of with the POEs," said Hyde. "There aren't many lessons to be learned from one building, but instead taking buildings as a group and noticing the trends. We hope to be the recognized market experts in post-occupancy evaluations."

Deliver More Than You Promise

Approaching 2005 – and reflecting on the changes of the last ten years since the C.K. Choi Building project – Kevin Hydes has likened KEEN's "green journey" to turning an oil tanker:

> It can take six or eight miles for an oil tanker to change its course...through a whole series of subsequent, successive steps. In looking back now, we can trace what happened to us as a firm. But I personally changed, too, after meeting Bob Berkebile ten years ago – I'll tell that to anybody, anytime. He just flicked a switch that changed the whole context for me of what it means to be an engineer.

In looking ahead, Hydes prefers to learn from the past. He remembers a lesson from one of the many management books he read in the 1980s: "deliver more than you promise."

> I think that phrase matches up well with our green arrival. In our commercial work during the '80s, we were always responsive. We asked questions and listened to the answers – but the answer to what question? To be engaged, instinctive, committed and enthusiastic – that's how green fits into KEEN today. It definitely doesn't come easy. It starts from within. "KEEN" is an adjective.

Calgary Water Centre

Location: Calgary, Alberta

Type: Office Building

Size: 156,082 sq. ft.

Completion: Scheduled for Spring 2006

Rating: LEED® Registered

Architect: Manasc Isaac Architects, LTD., Sturgess Architecture

Sustainable Features:

- Raised floors, underfloor air distribution
- In-slab radiant cooling
- 100 percent daylighting
- Over 50 percent less energy than National Energy Code
- Storm water storage for irrigation
- Capturing water for flushing toilets
- Dual-flush toilets, low-flow fixtures
- Indirect evaporative cooling

Photo courtesy of Strugess Architecture

Back to the Future

Becoming KEEN 3

The next business frontier is rethinking everything we consume: what it does, where it comes from, where it goes, and how we can keep on getting its service from a net flow of very nearly nothing at all – but ideas.

– Paul Hawken, Amory Lovins and L. Hunter Lovins,
Natural Capitalism

Great engineering can be *replication*, which is what Henry Ford did. Or it can be *invention*, as Thomas Edison showed us. So we'll need to continue to include innovation in every piece of engineering we do from now on, because that's what makes KEEN different.

Certainly, we've got to deliver good customer service. We've got to deliver quality. And we've got to be able to provide innovation. The real job of everyone in this room is deciding what that innovation looks like on behalf of the client: do they want invention or replication or some combination of both?

With that, Kevin Hydes kicked off KEEN Engineering's 2005 Spring Management Conference in Vancouver in late February – three days devoted to cross-company planning, updates, strategy and celebration. It was a time to reflect on KEEN's forty-five year tradition and a time to look introspectively at its strengths and weaknesses and at what the future could be.

Innovation: Replication and Invention

One glimpse of that future – perhaps the most visible and certainly the largest KEEN project ever – is already underway in Vancouver. Construction began in November 2004 on the $565 million expansion of the Vancouver Convention and Exhibition Centre (VCEC). Extending out over the shoreline of Vancouver's Coal Harbour, the facility will accommodate the needs of larger national and international trade shows and conferences, also serving as the broadcast center for the upcoming 2010 Olympic and Paralympic Winter Games.

Today, postcards of downtown Vancouver often feature images of the immense white sails over the current convention center – a feature that will remain in place along the harbor, soon to be joined by a green, landscaped roof over top of the new space; in fact, it will be the largest living roof in North America. This element is one of many elements included to meet the city's sustainability mandate for the additional 359,000 square-foot facility.

For KEEN, retained as the project's mechanical, plumbing and fire protection engineer and much of the electrical design, the VCEC is a perfect opportunity for its engineers to replicate and enhance the innovative designs they have created for earlier projects. In collaboration with the project architect, Musson Cattell Mackey Partnership, KEEN's strategies include green roofs, natural ventilation, on-site wastewater treatment and radiant heating and cooling – this time on a much larger scale.

"In most cases, we're just applying the systems we know how to do – and do well," said Hydes. "This building in itself will probably save more total greenhouse gas emissions through these sustainable strategies than maybe twenty or thirty past KEEN buildings that in many ways are more progressive."

> Look at the world around you. It may seem like an immovable, implacable place. It is not. With the slightest push – in just the right place – it can be tipped.
>
> – Malcolm Gladwell
> *The Tipping Point: How Little Things Can Make a Big Difference*

Other sustainable features for the Convention Centre will include an advanced blackwater treatment plant, using micro-fiber filtration that results in very high-quality water for irrigating the green roof and an adjacent park in summer and for flushing toilets and urinals. The majority of heating throughout the expanded space will come from chillers used as heat pumps; in turn, seawater will serve as condenser water for the chillers, eliminating the need for cooling towers and dramatically increasing system operating efficiency. Inside, the new 250,000 square-foot Exhibition Hall will incorporate radiant cooling in the roof structure; in-slab floor radiant systems within meeting rooms; and ballrooms and pre-function areas will absorb the solar loads of large perimeter glass areas in summer and offset cooling effects in winter.

"Why work on these large-scale projects?" asked Hydes. "To me, it's all about making them as green as they can be. It's about leverage. And it's about greenhouse gases."

Today, more than a decade after KEEN Engineering's transition to green first began, and five years since it declared a company-wide commitment to sustainable design and consulting, the LEED® standard has unquestionably transformed the market while redefining the parameters of occupant comfort and energy efficiency across North America. Still, USGBC's goal to impact the top 25 percent of new and existing buildings raises some questions for Hydes and others at KEEN, a firm clearly dedicated to living and promoting LEED® – what about the rest of the market? KEEN, therefore, is determined to concentrate on three distinctly different strategies going forward; first, on architects, owners and others driven to go beyond LEED® standards in their buildings; next, on the primary LEED® audience itself – to learn, to motivate and accelerate the system that already exists; and, finally, on representatives of the business community, institutions and government who believe that LEED® certification represents only a specialty niche or is too expensive.

And the World is Watching

The City of Vancouver was selected as the host city of the XXI Olympic Winter Games to be held in February 2010. Among its premier venues will be the Richmond Olympic Oval, home to a state-of-the-art, long-track speed skating venue with seating for eight thousand spectators. Located along the banks of the Fraser River and the centerpiece of a $155 million complex, the Oval will be a showcase multi-purpose sports, recreation and community facility both before and after the Games. The Richmond Olympic Oval is scheduled to open in late 2007 just in time for the national speed skating competitions in 2008, then the world championships in 2009, and finally the 2010 Olympic Winter Games.

In late 2004, KEEN was retained by the Vancouver Olympic Games Organizing Committee to handle the mechanical and electrical design, joining Cannon Design, a U.S.-based architect and leader in recreational/leisure/sport facilities internationally. The Vancouver Games also represents a conscious effort to broaden and strengthen the focus of sustainability for the Olympic Games, building on a decision made by the International Olympic Committee (IOC) more than a decade ago to environment as a third pillar of the Olympic Movement.

"The Oval is undoubtedly the gem of the new facilities being built for the Olympics," noted Hydes. "It's also one of the most significant projects in our firm's history and will be seen by a global audience, by more people than any other building we've ever worked on. It's going to be a dramatic structure – and elegant in terms of engineering – that will house the venue that's always one of the most intense and exciting competitions of the Games." "For us to be working on it and publicly displaying it as a sustainable project," he added, "offers so many opportunities to get the word out about green buildings."

KEEN is considering a wide range of renewable energy, wastewater treatment, air quality and other strategies for the Oval. As lead designer on the project, Blair McCarry and his team are also actively investigating the use of several new technologies, some piloted on the many smaller community and recreational facilities KEEN completes each year. These technologies include daylighting in large volume spaces without affecting the ice and using new heat recovery techniques to create new ice for the skating surface.

For Canada and within the global context, the sustainable goals of the Vancouver Olympics – in terms of reducing greenhouse gas (GHG) emissions – line up perfectly with the country's broader commitment to the Kyoto Accord, which concludes in 2012.

February 16, 2005 was an important date for Canada and KEEN Engineering – the day the Kyoto Protocol commenced. To mark the occasion, KEEN held its "Kyoto Gala" at the Vancouver Pan Pacific

Vancouver Convention and Exhibition Centre

Location: Vancouver, British Columbia

Type: Convention/Meeting Facility Expansion

Size: 359,000 sq.ft.

Completion: Under construction, completion expected in summer 2008

Rating: LEED® Registered

Architect: Musson Cattell Mackey Partnership and LMN Architects

Sustainable Features:

- Largest green roof in North America
- Micro-fiber filtration blackwater treatment system for irrigation, toilet flushing and urinals
- Seawater as condenser water for chillers
- In-slab ceiling and floor radiant heating and cooling
- Fire protection systems feature seawater source back-up

Copyright © LMN Architects

Hotel to acknowledge Canada's commitment to the Kyoto Protocol and to celebrate KEEN's 45th anniversary as a firm. Canada's goal under the Kyoto Protocol is to reduce its GHG emissions to 6 percent below 1990 levels by the year 2012.

During the Gala, KEEN staff mingled with invited guests – city officials, local architects and planners, university representatives and the press – surrounded by banners and detailed informational posters about Kyoto, climate change and GHG statistics. KEEN staff members were able to discuss the Protocol and answer questions about its implications for Vancouver and Canada. "I encouraged people to talk with clients and explain that all their buildings could be 'twice as good' buildings through sustainable design and improved efficiencies," said Rosie Hyde, KEEN's Director of Research who was charged with briefing all staff prior to the event.

Just a year earlier, the Canadian government officially launched its "One-Tonne Challenge" campaign, calling on all individual Canadians and businesses to reduce their emissions by one tonne – about 20 percent – through lower energy usage and other lifestyle changes. KEEN took that challenge a step further, issuing a challenge to everyone on its own staff – as well as engineers across the industry – to reduce their greenhouse gas emissions by 1,000 tonnes on an annual basis through better design. Featured in *Canadian Consulting Engineer* recently, Hydes and Hyde observed that:

> One engineer designing exclusively "twice-as-good" buildings would be able, every year, to reduce Canada's total annual greenhouse gas emissions by more than 1,000 tonnes CO_2, for a 26,000 tonne reduction of annual emissions by 2030. Multiplying annual savings by a fifty year service life for buildings, that engineer's work saves the planet more than 1.3 million tonnes....The ongoing greenhouse gas emissions associated with the operation of these buildings, year after year for the next half-century, will be part of the legacy we leave to our children.[1]

The North American Green Leader

KEEN believes it has earned the right to be called the "North American Leader in Sustainable Engineering" – an achievement that KEEN employees take seriously and one that consistently informs their actions as a firm. Creating change and assuming a leadership role in North America's building industry is tightly woven into KEEN's mission as a firm.

> The future is not some place we are going to but one we are creating. The paths to it are not found but made, and the activity of making them changes both the maker and the destination.
>
> – John Schaar, futurist

In spring 2005, Hydes, KEEN's President and CEO, became the new chair for the U.S. Green Building Council (USGBC) and its more than 5,500 member firms. An early participant in the Council and co-founder in 2003 of the Canadian Green Building Council, Hydes saw his new role as an opportunity to help broaden the reach and impact of sustainable design across the continent. To that discussion, he posed the questions: what about the mass market, and what about the "beyond LEED® " thinkers who are advancing the art and science of green buildings to new potentials?

Hydes says that the USGBC must continue to target the leading 25 percent of the building projects; but some estimates are now showing LEED® 's impact as just 1 percent to 5 percent of the market after the first five years:

> The challenge for me currently, as USGBC Chair, is to make sure we get the other voices at the table as well – that we hear from the business community and the mainstream developers who continue to build hundreds of millions of square feet of new buildings every year. Why isn't every project that they do LEED® certified or a green building? Imagine the impact of 2,000 architects designing LEED® buildings...

I think there's an implied requirement right now in the marketplace to do more than just a code-minimum building. The market is advanced enough and our experience is sufficient at this time to know that we can just do smarter things that do have environmental benefits without creating any economic downside.

KEEN's evolution, growth and ability to remake itself as a business has followed its own distinctive "bell curve". It is a symbol Hydes uses to describe the firm's past and present – "KEEN 1", "KEEN 2" – and its ongoing transition to becoming "KEEN 3". KEEN 1 represented the mass market, delivering innovation on conventional projects but primarily as a traditional engineering firm. Becoming KEEN 2 represented a new way of thinking, through projects like the C.K. Choi Building and Library Square and the creation of KEEN Concepts and other new services. Now, KEEN 3 means continually expanding the firm's focus: back to the mass market, applying the green principles it has already learned, and ahead to "beyond LEED® " projects, sharing those new ideas and techniques with the rest of the market.

To Hydes, green should be standard on every project:

When asked what percentage of our work involves sustainable design, I've often waffled on the answer as being around 50 percent. The more I think about it, that's complete nonsense: it's got to be 100 percent. Why would we choose not to do something green on every job? It's almost unethical. If we know we can do something and we know there's no risk to it and no cost to it, then we should just do it.

Further reinforcing KEEN's drive to change the market – and its own ongoing internal transformation – is the recent commitment to ISO 14000 standards on environmental management. ISO 14000 is a structured approach to addressing an organization's environmental bottom line, with specific guidance on how to evaluate and improve environmental performance. These standards are also designed to be compatible with quality management standards.

Richmond Olympic Oval

Location:	Richmond, British Columbia (Greater Vancouver)
Type:	Recreational Facility; Speed-Skating Venue for 2010 Olympic Winter Games
Size:	350,000 sq.ft.
Completion:	Under construction, completion date of Spring 2008
Rating:	LEED® Registered
Architect:	Cannon Design

Copyright © Cannon Design

In 2005, KEEN Vancouver began applying ISO 14000 to all MEP and business processes used among the office's ninety engineers, adopting the best of its green practices for all new jobs. Further, KEEN is auditing every one of its approximately one thousand projects annually to assess the actual impact of those design decisions: on water and energy efficiency, occupant comfort, operating cost savings and other improvements that collectively exceed the performance of building codes year after year.

"So if we can save fifty tons of greenhouse gases on a building because of the improvements we've made, then we need to start totaling the impact of that on an annual basis," noted Hydes. "I'm predicting that five years from now, it will be a requirement to declare these types of numbers. So it's important for us to start to get this new thinking within our firm. We want to be able to say to owners and architects: 'This is the impact that we think we're seeing with these sustainable approaches.'"

A Prototype for Green Living

Separated from Vancouver by the Strait of Georgia and from Washington State by the Strait of Juan de Fuca is Victoria, the capital city of British Columbia. World-renowned for its picturesque Inner Harbour, heritage buildings and colorful gardens, Victoria is considered one of the most livable cities in North America. The old industrial Dockside Lands located downtown, however, have long posed a challenge for city leaders: the brownfields site along the waterfront is made up mostly of contaminated landfill containing garbage and petrochemicals.

That blighted, forgotten fifteen-acre landscape is about to change, the result of a new $300 million, mixed-used development that may become a prototype for sustainable communities in the future. Known as Dockside Green, the site will be the biggest development of city land in Victoria's history, planned as 1.3 million square feet of residential, office, retail and light-industrial space. The city, in partnership with the Windmill Development Group and VanCity Enterprises, has mandated that all buildings on site be certified LEED® Platinum.

For KEEN, too, Dockside Green represents an exciting new piece of the future: not simply the idea of brownfields redevelopment, but an opportunity to participate in an advanced green project that represents a complete, developed package. Two years ago, the firm was invited to join the development team that included Joe Van Belleghem of Windmill, a Victoria-based pioneer in LEED® buildings, and Peter Busby of Busby Perkins + Will, along with VanCity as the financial partner. In December 2004, the City of Victoria selected Windmill's proposal – largely on the merits of its innovative environmental features.

"There's going to be a lot of magic happening with Dockside Green," said Blair McCarry, the lead designer for KEEN. "Its real significance is that the entire concept is based on an assessment of the 'triple bottom line.' This development is taking that idea to a new level: creating new jobs on site, revitalizing the local economy with viable businesses, and aiming for a greenhouse gas-neutral community that's also a net producer of clean energy."

Dockside Green's master plan calls for 860 units of housing, a small hotel, a commercial village and 75,000 square feet of offices and lofts. The Wise Energy Co-op is also planning to locate a bio-diesel factory on site to produce significant amounts of fuel per year. Various demonstration uses for the bio-diesel by-product – made from kitchen wastes and restaurant grease – include fuel for the harbor ferry, mini-transit vehicles and Dockside's car-share smart cars.

In addition to serving as a sustainability adviser for the development, KEEN will be responsible for all mechanical and electrical design. The development's main goal is to generate sufficient renewable energy on site to meet its entire heating, cooling and electrical needs. Central to this scheme is a wood waste-to-energy utility plant – biomass energy co-generation – that will produce approximately 2 MW of electricity and 3 MW of heat output. Wood waste is typically available from nearby mills, woodworking shops, and the tree trimming and deadfall in the CRD. Also included will be geothermal heat pumps for commercial buildings where cooling is required.

The plans for water and wastewater are no less innovative. All sewage will be treated on site, utilizing a sophisticated bioreactor tank/ultra-filter membrane system to process all blackwater to near drinking water quality for irrigation on Dockside's green roofs, for landscaped water features and ponds, and for toilet flushing. Extensive water conservation measures will be employed throughout the development, dramatically reducing potable water use in buildings. In addition, stormwater collection and storage systems will supplement the treated blackwater for irrigation and to feed creeks and ponds along a greenway running the length of the property.

"Dockside Green has the potential to create a unique international dialogue about what sustainable communities can and should be", said Hydes. "When we consider what the future is all about in terms of this triple bottom line – socially, economically, environmentally – how do we describe the buildings we're building, how do we describe their efficiencies, and how do we program in long-term efficiencies over time? That's really where we need to reach with a very visible project like this."

Embracing Change, Sustaining Change

Influencing an issue, influencing codes, influencing the MEP industry – even the design practices of direct competitors – and the architectural profession, remains a core tenet for KEEN Engineering in greening North America's buildings. Whether active involvement in USGBC or speaking at conferences and to the consulting engineering community, the firm maintains a deep, overriding belief in change.

"Certainly, we got out early on this issue in some markets," said

Dockside Green

Location: Victoria, British Columbia

Type: Mixed-use Residential/Commercial/Industrial Development

Size: 1.3 million sq.ft.

Completion: Under construction, first buildings scheduled for completion in January 2007

Rating: LEED® Platinum Expected

Architect: Busby Perkins + Will

Sustainable Features:

- Greenhouse gas-free energy for electricity and heating
- Biomass energy co-generation (through gasification of wood waste) to generate district heat and electricity
- On-site sewage treatment facility utilizing bioreactor tanks and ultra-filter membranes to treat all blackwater to near drinking water quality
- Bio-diesel factory, producing fuel for harbor ferry, mini-transit and car-share smart cars
- Geothermal heat pumps for commercial buildings where cooling is required
- Reduced potable water use
- Stormwater and treated blackwater for irrigation on green roofs, water features and ponds

Hydes. "What I'm trying to do in the professional arena really is talk about KEEN's business and engineering success relative to taking a very strong position on sustainability – you know, literally describe what we've done in detail, give engineers the 'how to' in tapping this market to further transform the market."

One recent KEEN-authored article offered insights on setting fees to ensure profitable green design projects that can generate additional revenue and income. In the April 2005 issue of the *ASHRAE Journal*, Tim McGinn of KEEN's Calgary office contributed his experience with a high-end corporate classroom/conference center on a community college campus with a budget of $14.5 million:

> Fee arrangements for green building projects recognize the shift of the mechanical engineer's role as part of an integrated design team. This higher degree of integration during the conceptual design stage demands active involvement from all participants. The challenge for mechanical design firms is to maintain full participation from the architect, owner, and others and still remain profitable.
>
> As the green building industry rapidly moves towards the mainstream of design, the challenge for firms not currently involved in designing green buildings is to adapt to maintain market share. Firms need to gain the experience and expertise to efficiently design green buildings. A consultant with a demonstrated expertise in green design often will receive the commission based on the value brought to the team.[2]

"I think what we've been able to demonstrate to the industry is that green design isn't a passing fancy, it's a mainstream issue," said Hydes. Ultimately, KEEN realizes that this approach gives new meaning to the design process, particularly for the next generation of engineers, and represents an emerging new frontier in mechanical and electrical systems thinking. That is also part of the KEEN story going forward: the drive to attract and retain top talent from across the U.S. and Canada – those who are interested in non-traditional design assignments and in learning the leading-edge core competencies required of green engineering. "Today, our sustainable position creates a lot more clarity around what KEEN is all about," added Hydes. "So there's a lot more alignment and self-selection among our candidates even before they come to the interview table. As a result, we're more likely to be successful in hiring, and we stand a far better chance of making the type of impact we're trying to make."

One metric tonne equals 1,000 kilograms. The volume of one tonne of greenhouse gas emissions (GHGs) would fill an average two-story, three-bedroom house.

In addition, KEEN takes a proactive stance on teaching, bringing its message of sustainable design and technical proficiency to university graduate and undergraduate students. No fewer than five KEEN principals now volunteer as university-level guest lecturers or adjunct professors in Seattle, Vancouver, Calgary, Toronto and Montreal. Others serve on curriculum committees for engineering and architecture schools.

"Again, there's a self-interest in helping these programs so that every engineering graduate coming out of every school has a basic palette of understanding of environmental issues," said Hydes. "Traditionally, there's been very little training on this knowledge among our rank-and-file engineering graduates, so we're trying to go right upstream of the problem and influence the curricula."

Tomorrow... Just Arrived

It is quite possible – even likely – that some of those engineering graduates will work on what could be the most innovative, high-performance building in North America: CIRS, the Centre for Interactive Research on Sustainability. To be located near downtown Vancouver, the CIRS concept was created by the University of British Columbia to accelerate the adoption of sustainable building and urban development practices in response to the immense challenges facing cities as a result of population growth and environmental degradation worldwide. The $15 million center will foster sustainable research and design and demonstrate advanced information technology, computer modeling and advanced visualization capabilities.

Professor Ray Cole of UBC's School of Architecture recently described the scope and vision of CIRS in *Canadian Architect*:

> Green building practices continue to mature and find widespread adoption through incremental improvements in performance efficiencies. While an important first step, simply producing buildings that are progressively better than typical practice will prove insufficient to meet the requirements of a built environment that can support sustainable patterns of living within a context of rapid urban development. Greater performance leaps will be necessary and at a faster rate. This will challenge many existing norms and expectations and, in particular, redefine how we conceive the design, construction and operation of buildings.[3]

As part of the new Great Northern Way campus – essentially a "sustainability precinct" in Vancouver – CIRS will accommodate four academic institutions, including UBC, with space for researchers to collaborate on interdisciplinary and inter-institutional projects. Partners from the private, public and NGO sectors, such as the David Suzuki Foundation, will also share the research facility.

Following the project's initial feasibility study in late 2001, Busby Perkins + Will Architects developed the initial concept for the nearly 130,000-square-foot research facility, which has three main components: two institutional wings and a tenant wing separated by two central atriums. As the mechanical and electrical consultant, KEEN began schematic design in June 2005. Construction will commence in summer 2006, with project completion in late 2007 or early 2008.

Among the chief design goals for this off-the-grid green building is being greenhouse gas neutral and being a net annual green power producer. In addition to incorporating energy-efficient building products, CIRS' passive design, including 100 percent daylighting and ground-source heat pumps, will result in very low energy requirements. Rainwater will meet potable water requirements, all wastewater will be purified using a bio-filter system, and remaining stormwater will be controlled, re-used and discharged on site. A living roof, photovoltaics and micro-hydropower are a few of the other green features under consideration.

Equally significant, the CIRS structure itself will be used as an ongoing, state-of-the-art "living laboratory". Over the lifetime of the building, CIRS researchers and building industry partners will monitor, analyze and document hundreds of data points to assess the effectiveness of new and existing sustainable building technologies, as well as the interaction of building occupants with these systems.

KEEN Vision

Where CIRS, Dockside Green, and other recent projects signify the firm's clear intent to explore new sustainable concepts and technologies, KEEN senior management – Mitchell, McCarry, Hydes and Anseeuw – also recognize the importance of leaping even further ahead to the next generation of ideas. Beyond LEED®. KEEN 3.

That leap, said Hydes, is "KEEN Vision", which would build on the foundation of KEEN's long tradition of excellence as a firm and, more recently, the creation of KEEN Concepts group as its green R&D team:

> I think what I'm talking about is a re-invention, a re-launch, but an intentional one, that may ask some very fundamental questions about the nature of design today or the direction we're headed in as a firm.

Centre for Interactive Research on Sustainability (CIRS)

Location:	Vancouver, British Columbia
Type:	Sustainability Precinct with Educational/Office/ Retail Space
Size:	60,000 sq.ft. (Phase 1)
Completion:	Late 2007/early 2008
Rating:	LEED® Platinum Expected
Architect:	Busby Perkins + Will

Sustainable Features:

- Greenhouse gas neutral facility – net annual power producer using sustainable and renewable energy sources, including photovoltaics and micro-hydro
- 82 percent of Model Energy Code
- 100 percent daylighting as primary daytime light source
- Full blackwater treatment
- Living roof
- Ground-source heat pump system for cooling and heating
- High-performance exterior glazing
- On-site rainwater collection and re-use

Copyright © Busby Perkins + Will

Becoming KEEN 2, for example, gave us something else we could talk about that had meaning in terms of the social, economic and environmental benefits of green design. Yet there's always a danger that once you start to institutionalize certain beliefs and practices, like anything else, you may start to limit your work within known boundaries. So we need to continually challenge what we've learned, to self-examine, and to measure everything against the idea of First Principles. I definitely feel at this point that we've got to make the next leap to this KEEN Vision world.

I see a world where LEED® is not an option – it's required, it's the minimum level of performance in constructing, renovating or upgrading any building. After all, if we can incorporate green elements into every job we do, save energy, with no risk and without spending a dollar more of the client's money, why wouldn't we do it? It's simply a matter of being a good engineer, a good citizen. And it's the right thing to do.

Section Two

The Principles

The Principles

'Think simples' as my old master used to say - meaning reduce the whole of its parts into the simplest terms, getting back to first principles.

– Frank Lloyd Wright

Introduction – The idea of the "Ecological Engineer" is centered on the notion that engineers have a wider responsibility to the natural environment than most individuals in virtually any field. In fact, because of the sheer size of the environmental footprint associated with the built environment – of which engineers are critical in shaping – this responsibility must inform every action that an engineer takes with the goal to provide engineered systems that are highly functional and effective while minimizing or eliminating negative impacts to the environment. How an engineer approaches his or her craft, and the fundamental ideas that shape it, is what sets the ecological engineer apart from his or her peers.

KEEN Engineering was selected as the first "ecological engineering" practice to be highlighted in this series, in large part because of the very strong and clear guiding principles that inform the firm's direction and practice. At the heart of KEEN's practice is an emerging passion for sustainable design that has been documented in the first section of this book. Employees at all levels of KEEN see themselves as not only engineers, but as champions of an ethic that has the potential to shape societies' core relationships with their buildings and the larger environmental context.

This chapter explores the major philosophical principles that shape the best ideas emerging from KEEN in their offices all over North America. They provide a model for engineers of all backgrounds – indeed of any building professional – on how to guide critical decisions for any building project.

Principle One – Returning to the Basics

Understanding the "First" Principles of Engineering

In many ways conventional engineering practice has become a discipline driven by "rules of thumb" and "standard practices" that see similar – and "proven" solutions appearing time and time again. In many cases, these solutions are not derived from the careful analysis of a specific project's needs, but from slightly modified versions that "worked well enough" on the last project, which is sometimes defined as solutions that produced few complaints. At other times engineers rely too closely on minimum "code" standards that are routinely behind the "best practice" curve by a significant margin. Energy efficiency, or environmental performance in general, is often not a feature of these conventional and routinely repeated solutions. An institutionalized mediocrity of "safe solutions" has crept into the profession.

Part of the problem is that, in general, engineers are not rewarded for efficiency, but sufficiency. People do not complain much if energy bills are slightly high, but they do complain if a system is improperly sized and comfort conditions cannot be met. Low fees and linear design processes, where the engineer is brought in too late for innovation to really occur, exacerbate the problem. Sometimes even the fee structures work against a more ideal process – the more innovative the engineer becomes, the less he or she is rewarded. Fees should always be based on performance, not size.

The "Ecological Engineer," like KEEN, realizes that solutions can be both efficient and economically effective – but they require that the engineer take a critical look at the specifics of any project, its climate and functional needs. Rule-of-thumb solutions are not instantly trusted, but instead are analyzed using fundamental lessons of engineering – many lessons learned in school, but often forgotten. Having a firm grasp of the principles and physics of engineering such as heat transfer, fluid dynamics and human comfort are essential. These "first principles" should be used to guide the start of any project and used by engineers to check their assumptions. For example, even using energy modeling software is not enough to ensure that the best solutions are being pursued as modeling can be used to institutionalize certain biases and justify pre-determined results.

At KEEN, it is expected that their engineers will check their own assumptions at the door and view design assignments with a fresh set of eyes – even if it might be the same building type as another project recently completed. For example, are the assumptions still valid? Can the previous solutions be improved? What were the barriers to greater efficiency on the last project? How does climate, micro-climate, and site resources allow for new solutions that might be more innovative?

While energy modeling and other computer simulations are used at KEEN, they are not taken as gospel – engineers are encouraged to check results with hand calculations – and to really think about the basic physics behind the answers – does it make sense? Is that how air would really behave in a space? Do the answers compare to reality? KEEN also understands that sometimes solutions exist because of history – the way things have always been done – and their engineers are encouraged to look to other countries and how they solve problems. Europe and Asia often take an entirely different approach to solving similar problems in similar climates as here in North America. Looking at things from their perspective tends to remove institutionalized biases. Examples also abound from other industries. Understanding basic engineering principles at a fundamental level allows the ecological engineer to think outside the box and source innovative solutions from unlikely places.

An example of this principle in action can be found in KEEN's increasing use of condensing boilers that are more common in Europe, instead of more conventional boilers typically found in the United States and Canada. Condensing boilers can be up to 95 percent efficient due to the process of condensing flue gases and capturing their heat instead of letting it escape out of the system. Since the condensing gases

In looking outside of North America, KEEN has borrowed a number of "ecological engineering" solutions from Europe and Asia - such as less-polluting condensing boilers that can be up to 95 percent efficient by condensing gases and capturing the heat.

are typically water, this equipment has the added benefit of lower pollution emissions, critical for the ecological engineer to consider.

KEEN works hard, through innovative coil selections and system designs, to ensure that these condensing boilers deliver as well at part load as they do at full load – ensuring greater savings and bucking conventional practice which would shy away from reducing return temperatures so much. Returning to "first principles" opened the door to this possibility.

Principle Two – Remembering the Environmental Mantra

Reduce, Reuse, Recycle and … Rethink

During the eighties and nineties a slogan was developed for the environmental movement that was affectionately nicknamed the "Three R's." The three R's stood for Reduce, Reuse and Recycle. The slogan implied action towards conserving precious natural resources as opposed to the dominant paradigm of consumption, more consumption and even more consumption. But even more important than the words themselves was the deliberate order in which they appeared. First Reduce – which asked individuals to ask for less rather than always more. Then Reuse – which asked people to consider extending the life of a pre-existing object before replacing it and then finally, Recycling – which suggested an appropriate end-of-life for a material or technology other than simply the landfill. While most of the public skipped over the first two principles, recycling as an accepted idea has blossomed and recycled-content options are available for many consumer products in the marketplace.

For the ecological engineer, it is important to realize that this environmental mantra does not just apply to consumer goods, but can also serve as an operating philosophy for engineering practice. Engineers should think about problem-solving using a pre-determined order of environmental benefit, with the goal of minimizing environmental footprint effectively. At KEEN, engineers are encouraged to use this philosophy to question their own solutions, and yet they go an important step further by adding a fourth principle to the front end – Rethink.

The idea behind Rethink is first about challenging the fundamental assumptions of any project. At the C.K. Choi building in Vancouver, British Columbia the question was: "Does the building really need a mechanical system for comfort to be achieved?" – rather than assuming that it does and merely trying to reduce the size of a conventional system. On the Choi building, this Rethink opened up the option to naturally ventilate the building instead of relying on mechanical cooling. "Does the building really need a sewer connection?" was another Rethink question asked on Choi. The result was that Choi uses composting toilets instead of conventional flush toilets, saving the university thousands of gallons of water per year.

Framing questions at the beginning of a project in such a holistic way opens the door to smarter and more robust solutions. Following the "Environmental Mantra" through the design process allows the innovation and environmental performance to be improved. Thinking hierarchically in this philosophical order rather than immediately jumping to size ducts, pipes, transformers, or any other engineered building component always results in greater efficiency, and often at reduced first cost. The ecological engineer is always encouraged to start with Rethink – as the decisions with the most impact on the environment are always related to what did not need to be built in the first place.

A notable example can be found at the new Kent Avenue Testing Facility in Vancouver. On this project, the design team pushed the boundary on the kinds of things that could be re-used by re-thinking the project possibilities at the very beginning and "borrowing materials and equipment" from the old facility that was scheduled for demolition. Both the engineers and the architects (Busby and Associates) followed this philosophy and the result is a pleasing building with an incredibly high amount of salvaged, or reused equipment. For example, "newer" rooftop units and 30 percent of the plumbing fixtures were salvaged from other demolished buildings, saving financial and environmental resources.

© Copyright: Jim Burns

At the Kent Avenue Testing Facility (Vancouver), project goals were simple: To rethink, reuse and recycle as much as possible. KEEN, Busby Perkins + Will and the rest of the project team were able to incorporate mostly recycled building materials, including the wood trusses, rooftop units and 30 percent of plumbing fixtures.

Principle Three – Embracing Architecture

The Integration of Architecture and Engineering

There is an adage in the building industry that architects and engineers do not get along. Like any adage, there is some truth in it, but it is not the whole truth. On most projects the relationship between architects and engineers is cordial, albeit lukewarm, and reflects the fact that significant cultural differences often exist between these two sides of the profession. At times it does seem that architects and engineers are two entirely different breeds, with differing approaches to communication, different agendas and methods of working. Without a basic feeling of trust, respect and appreciation for the respective roles, communication can break down and problems can creep into any building project.

Some architects pay too little attention to the engineered systems of their building designs and expect to just hand off their work to be "engineered" at the end of the design process. In many cases mechanical rooms are sized without much discussion and lighting, mechanical and plumbing systems are expected to "fit" within prescribed areas of a design more as an afterthought to other concerns such as building layout and aesthetics.

Some engineers are similarly culpable by taking little direct interest in the architectural design, even if given the chance, and by rarely stopping to appreciate the subtleties of the design, the spaces created and the potential for future occupants to interact with the facility. Some engineers are guilty of only wanting limited information – "show me where to put it"– so that they can simply pick off-the-shelf solutions that will fit into any design regardless of the direction the design is heading.

The result from both sets of attitudes is an incredible loss of potential, not only for more functional solutions, but for sizeable cost savings, better quality designs and significantly improved efficiencies. These attitudes often explain why some buildings are routinely uncomfortable and awkward to be in, while consuming great amounts of energy and resources.

The ecological engineer rejects these attitudes and wholeheartedly embraces architecture and the design process, because he or she realizes that unless the architectural process is understood and appreciated, then an optimized solution is not possible. Engineers such as KEEN learn not to fight architects and the architectural process, but rather actively seek to participate in it. They are excited by the opportunity to work with the most talented architects on challenging projects.

Ecological engineers jump at the chance to participate in charrettes, to contribute ideas and to give feedback – because these professionals realize that integrating their ideas, issues and concerns early brings considerable benefits down the line. They also understand that they add value to the process and they can make building designs more effective. They learn to see the architect, not as an adversary, or

© Copyright: Jim Burns

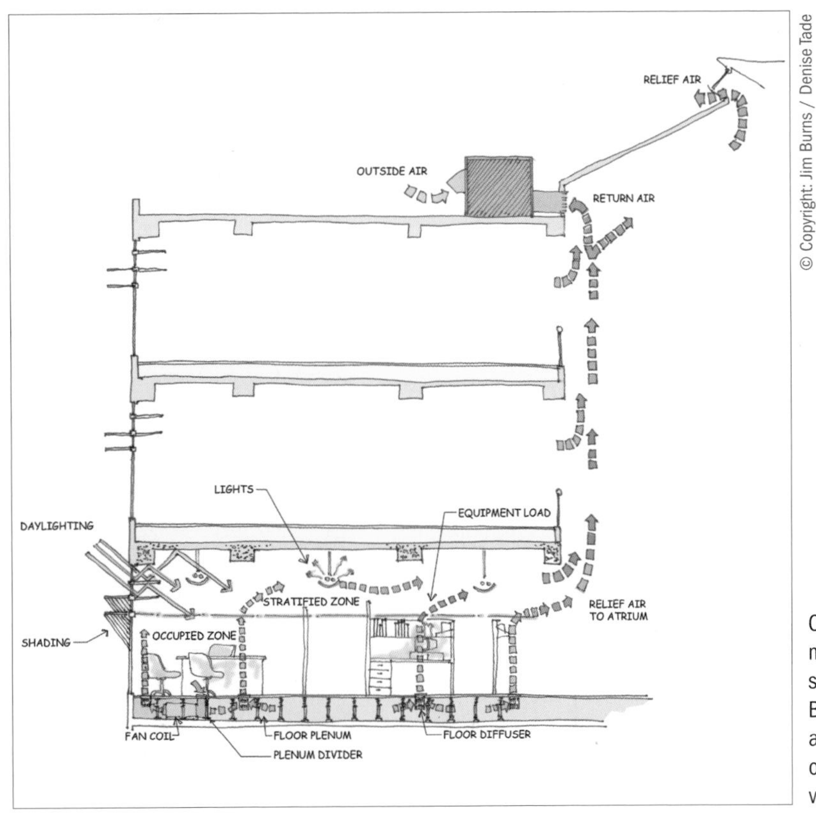

© Copyright: Jim Burns / Denise Tade

Collaborative solutions: KEEN's mechanical design helped to shape the BC Gas/Terasen Building (Surrey, BC) and its architectural features while optimizing underfloor air/natural ventilation and daylighting.

someone who is going to make their job more difficult, but as a co-collaborator in a process to produce the most beautiful, functional and sustainable building possible.

Ecological engineers also realize that architecture is a critical first part of the engineering solution and can not be separated out. For example, ecological lighting engineers realize that the whole building is a luminaire, and glazing, exterior fenestration and internal surfaces are merely the beginning stages of their own design process and technology selections. For ecological-minded mechanical engineers, it is the realization that comfort starts by first understanding the architectural design as it relates to the external climate conditions and then how it relates to the ways people use the spaces inside – thinking deeply about outside to inside and inside to outside.

Getting involved early affords the engineer an opportunity to help shape the architecture, its orientation, daylighting, materials usage and programmatic relationships – and thereby produce more elegant and efficient systems. Doing so also fosters a larger sense of ownership over the design of the building, which often translates into more thoughtful responses and further efficiencies.

For KEEN, the BC Gas/Terasen project was a perfect example of an engineering team embracing architecture and the architectural process. In this project, the entire shape of the building arose from intense collaboration with the architects (MCMP Architects) in order to achieve an optimized configuration. From shading to glazing selection these "architectural features" figured prominently into the thinking of KEEN's mechanical design team and, in turn, their requirements informed the building design and materials specifications. The balance of daylight with heat gain was incredibly important – to provide a well-lit interior without the burden of excessive cooling loads. Small, less expensive mechanical systems were the result of the team embracing architecture and the architectural process from the beginning.

Principle Four – From Components to Systems

Treating the Building as a Whole System – IslandWood

Often times the different disciplines that work on the design of a building approach their work in an isolated, linear fashion – coordinated by the architect yes, but operating in isolation somewhat. Systems are layered upon each other and "made to work" rather than starting from a perspective of integration. How various systems interact with each other is discussed only to the extent that there are no major conflicts such as between structural elements, lighting systems and mechanical systems. What does not happen often enough is a dialogue about how various systems can create synergies and positive interactions – in other words, how the building as a whole could be treated like a complete system for greater efficiency and clarity.

It is not possible to optimize a building's overall performance with each discipline working in isolation, or attempting to optimize their "jurisdictions" without thought of others. Each layer of the design – be it architectural, mechanical, structural, plumbing, electrical or otherwise – should be analyzed as part of a network of systems that make up a whole building. Sometimes synergies emerge that can save a great deal of money for the client; for example, when it is learned that waste heat from one process can be used to pre-heat another in an industrial application, or how selecting the right structural system can serve as thermal mass to lower temperature fluctuations and reduce heating and cooling loads.

The ecological engineer understands that every component of a building has the potential to affect everything else and therefore each component must be studied in some organized, systematic way whereby each discipline has an opportunity to present and discuss what is important to them. For example, solar shading and glazing specifications greatly affect heat gain and lighting levels inside a building as does interior color choices for lighting levels. Significant cost can be avoided by properly zoning a lighting solution to embrace daylighting and by properly downsizing mechanical systems to reflect the fact that there is now less internal heat gain associated with the electric lighting and solar gain. In a sense, the ecological engineer treats a building as an organism and tries to understand its whole behavior in order to seek solutions that are most beneficial. Some buildings, where this approach was not properly understood, end up being simultaneously heated and cooled, without thought for how waste heat could be directed to areas that need it.

For KEEN Engineering, an impressive example of this principle emerged on the IslandWood project, where architecture and engineering are closely intertwined in order to achieve a high level of performance(see the case study in section three). The design team members worked carefully together to shape the buildings' orientation and fenestration to meet goals that would allow for the buildings' mechanical system to not be utilized for the majority of the year as proscribed under LEED®

© Copyright: Jim Burns

Version 1.0. Each decision that affected heating and cooling loads – such as the amount of insulation and the type of glazing selection and exposure of thermal mass – was carefully considered and studied using energy models for possible interactions and synergies. Related goals for water efficiency meant that external surfaces of the building were examined carefully for water harvesting to offset potable needs within the facility and its grounds. Structure as thermal mass, roof tops for energy generation and water collection, and shading devices to reduce summer heat gain while reducing glare and improving lighting conditions are just a few examples found in this inspired project. IslandWood was made stronger by the design team treating the whole facility as a complete system, with architect and engineer working hand-in-hand from the beginning and seeking to understand how each decision supported or hindered the ultimate goals for sustainability.

Thinking in whole systems allowed KEEN and Mithun Architects to more fully integrate structural design and material selection with heating and cooling systems, rainwater harvesting and energy generation at IslandWood (Bainbridge Island, WA).

Principle Five – Going from Simple to Complex

Seeking Low-Tech Solutions Before High-Tech Solutions

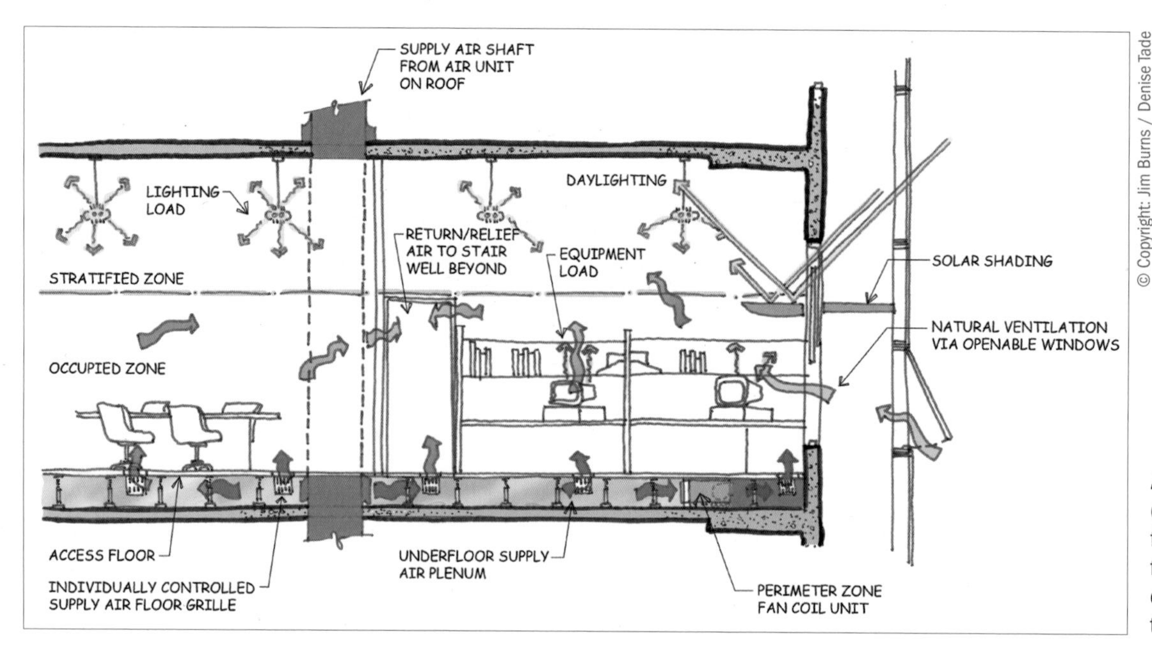

A concept sketch for the Telus Building (Vancouver), reflecting KEEN's practice of fully exploring 'low-tech', passive systems first as appropriate to climate and site conditions – and then considering higher-tech solutions.

For thousands of years human habitation was made "comfortable" only through the use of natural means – whether it be solar gain for heating or natural ventilation for cooling and daylight for lighting. Only in the last couple of hundred years has this changed radically as key inventions such as air conditioning and electric lighting made extending comfort to a year-round, night and day schedule possible. Indeed, it has only been in the last century or so that almost all natural means to provide comfort and light to buildings were abandoned in favor of completely mechanical and "high-tech" solutions. In many important ways this historical trend has resulted in more comfortable buildings with much greater operational flexibility and has had the effect of freeing up architectural form from restrictions imposed by natural methods of heating and cooling. Buildings could be built to look the same everywhere, regardless of climate, which became known as the "internationalism" of architecture.

The downside to this approach was not only a loss in regional differences, but in the incredibly large inputs of energy and resources needed to make this architectural form possible. The amount of energy used in conventional facilities is staggering (up to 60 percent of the world's electricity is used in buildings according to the US Green Building Council) and the majority of this electricity is produced through non-renewable sources such as coal, natural gas and nuclear power – all of which are having an extremely negative impact on the natural environment.

Perhaps even more disturbing is the attitude by many in the building industry that this energy draw is an acceptable trade-off for achieving modern standards for comfort and that there is little to be gained by understanding the lessons of the past. Indeed, there is almost a "cult of technology" existing, whereby some engineers seem to purposefully increase the use of technology for its own sake, even when more simple solutions exist, many of which they no longer trust or understand how to use. Often, passive or low-tech solutions have proven to be more effective than the best high-tech solutions available, at least for parts of the year and yet they are still ignored.

In contrast, ecological engineers realize that passive or low-tech solutions are a valuable part of their "toolkit" and should always be looked to first for solutions since the result is usually less energy, less cost and less maintenance. The ecological engineer also understands that passive systems, if properly designed, can result in superior levels of comfort or visual acuity than is possible with more high–tech solutions. By applying technical solutions only after lower-tech or passive solutions have been exhausted typically results in a better functioning building. In certain instances high-tech solutions can be avoided completely while still maintaining modern standards for comfort, as with some of the naturally-ventilated structures emerging in mild climates in the Pacific Northwest. The goal of the ecological engineer should always be to strive for the simplest low-tech solution possible.

A distinctive example of this philosophy is demonstrated by KEEN Engineering in the retrofit of the Telus building in Vancouver, British Columbia. The building needed a facelift to help revamp the image of the organization, and greater comfort and energy performance were desired. Instead of jumping to a conventional mechanical solution first, the design team suggested modifying the envelope to regulate external conditions while providing the new and more modern look for the facility. The result is a double envelope building whereby heat gain, glare and heat loss are regulated through the interstitial space between the two façades, providing free comfort for much of the season by dampening heat gain and loss to the interior and allowing for natural ventilation. The design of the façade itself is high-tech, but the mechanisms it relies on are primarily passive. Through understanding of the Fifth Principle, an exemplary project was realized.

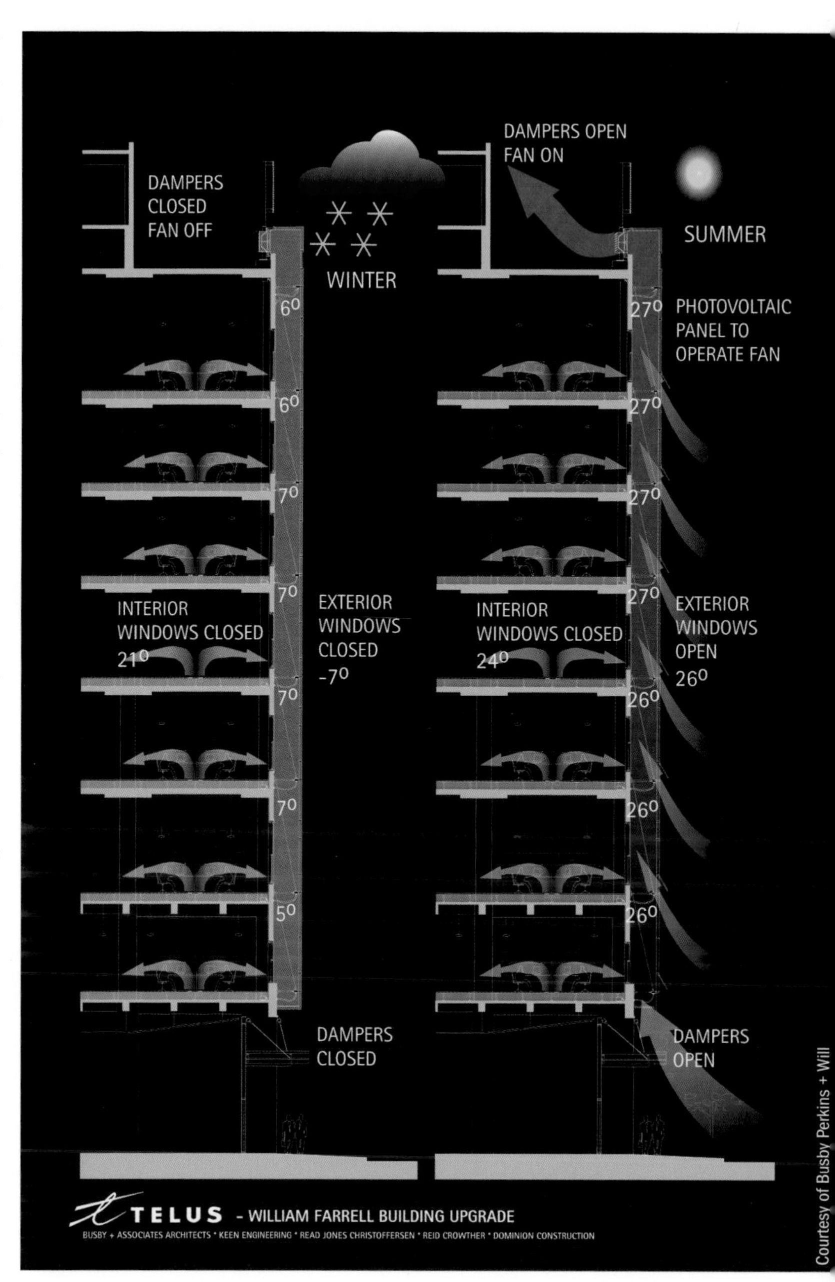

The Telus project, in winter and summer mode, resulted in one of the world's best naturally-ventilated buildings.

© Copyright: Jim Burns

The Telus revitalized: A new double-glazed, fritted and frameless glazing system with operable windows was suspended from the existing building face – keeping it simple and comfortable.

Principle Six – Getting Priorities Straight

Buildings for People First

In many modern buildings, despite the best technology that can be purchased, human comfort is still often wanting. People are either too hot or too cold and they have little ability to control their own thermal comfort. Building occupants are often at the mercy of the systems designed by mechanical engineers and they have to put up with rigid systems that do not respond to the myriad of conditions that exist each day in any facility. People are biological creatures – and everyone has a slightly different requirement for temperature and comfort. People dress differently, and different levels of activity exist within each job. Uniform comfort conditions ultimately result in many people being uncomfortable at any given time – the worst "sins" occurring when people are cold in air conditioned buildings in the summer, even when extremely hot outside or too hot inside when outside winter temperatures are frigid. Both examples point to a double failure of wasted energy and unsatisfied occupants.

There is sometimes an attitude in the engineering field whereby people are not "trusted" to handle their own comfort conditions because it is believed that they will "mess up the system" thereby resulting in increased energy use and more complaints. With this attitude, buildings are buttoned up and operable windows are removed. Centralized, automated systems regulate temperature and humidity without feedback from actual occupants – and facility managers often have inadequate knowledge of how to effectively override or properly manage these systems when it is desired. While there is some behavioral truth behind this attitude, it is usually the fault of the system design not being set up to allow for changes more than it is the occupants themselves. The "mistakes" made by building occupants to control their own environmental conditions quickly enter engineering folklore and become further rationale for removing control.

In contrast, the ecological engineer recognizes that the whole purpose of constructing buildings is to provide structures that work for people.

If occupants are not given a chance to interact directly with their own built environment then much is lost – including an appreciation for the systems that provide comfort to them. The ecological engineer recognizes that there is a responsibility to educate people on how their building operates and how to properly control their surroundings – whether it be operable windows, lighting controls or other means such as understanding set point temperatures. Even a small amount of control is often enough to produce a significant psychological advantage that results in people feeling more comfortable than before. System designs robust enough to handle these interventions become desirable. New technologies, like raised floor systems with under floor air delivery, make user interactions possible on an individual level as people can move air delivery locations and control how much air is moving around them. Allowing building users to have a more direct connection to comfort systems is far superior than waiting for complaints to drive changes in building operations.

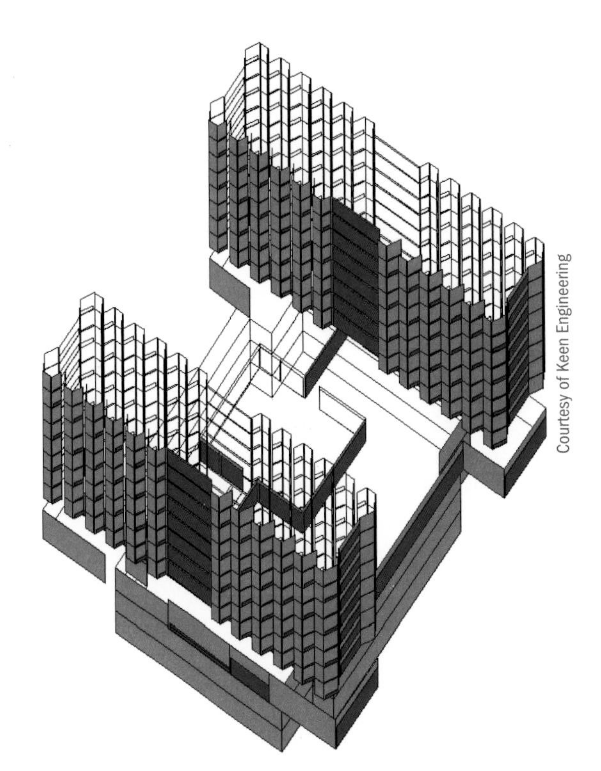

Modeling energy or natural ventilation dynamics is an important part of KEEN's "tool-kit," both in systems design and for educating clients and building users.

At KEEN, the engineers routinely remind themselves that they do not always know best. They begin by trying to understand how best to serve their clients – by asking them how they use their spaces and problems with comfort that they have had in the past. KEEN realizes that people cannot value what they do not understand or hear about, and communication regarding how their building works is extremely valuable. KEEN believes in creating operating manuals that can be understood by users and not just by other engineers.

A prime example of this principle in action occurred at the Child, Adolescent and Women's Mental Health Facility (CAWMHF) project where KEEN served as LEED® and Energy Modeling consultants. This building's primary function was as a mental health facility and the system needed to be incredibly responsive to the special needs of the population. KEEN's recommendations evolved through a series of key dialogues with the client and design teams. KEEN started by listening carefully to the client and by trying to understand what would work well for the client as a unique user group. The user group needed low-cost solutions, but the group members also knew that the proposed retrofits had to provide significantly improved comfort for their facility to work well for their functions. Allowing the occupants to interact directly with the building seemed to have direct merit, and yet the staff's previous experience had only been with buildings that were incredibly uncomfortable to be in.

While conventional cooling was first assumed, extensive natural ventilation modeling showed that it was possible to provide considerable personal comfort with minimal mechanical interventions as the existing facilities were upgraded and glazing and insulation were improved. A series of seminars was provided to demonstrate various options and to allow direct feedback from the users. KEEN's recommendations grew out of the process and, in the end, natural ventilation was adopted as a primary cooling strategy –saving energy, increasing comfort and allowing greater interaction from the building's occupants.

Principle Seven – Celebrating Engineering

Interjecting Fun and Visualization into the Practice

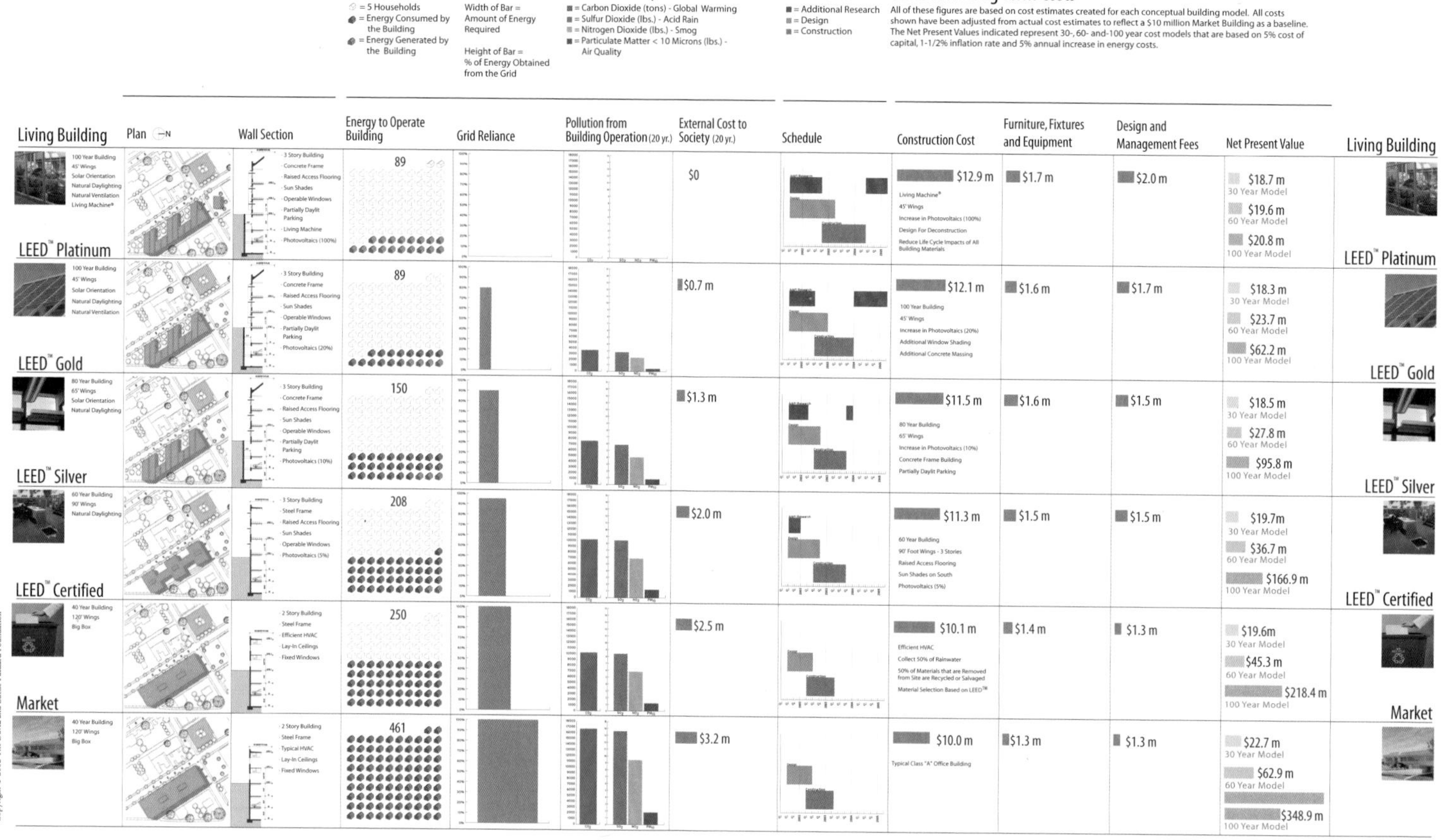

Engineering practice – be it structural, mechanical, electrical, plumbing or civil – often takes a back seat to the more flashy and tangible qualities of the architecture itself. Engineering tends to be an under-appreciated profession – the only time the engineer is paid attention to is when something goes wrong and complaints are made about how a facility is functioning. When a building is properly engineered, it tends to be at its most invisible, and the work, which should be celebrated, often is not.

This trend is made worse by the fact that the general public is often disconnected from how the world around them actually works – including their buildings and so they may not appreciate exemplary engineering practice even if they are staring right at it. The general public does not have a functioning language for describing their own thermal comfort, visual acuity and other needs. And yet, when things are carefully explained, most people find it interesting and empowering to understand how things that are very important to them actually work.

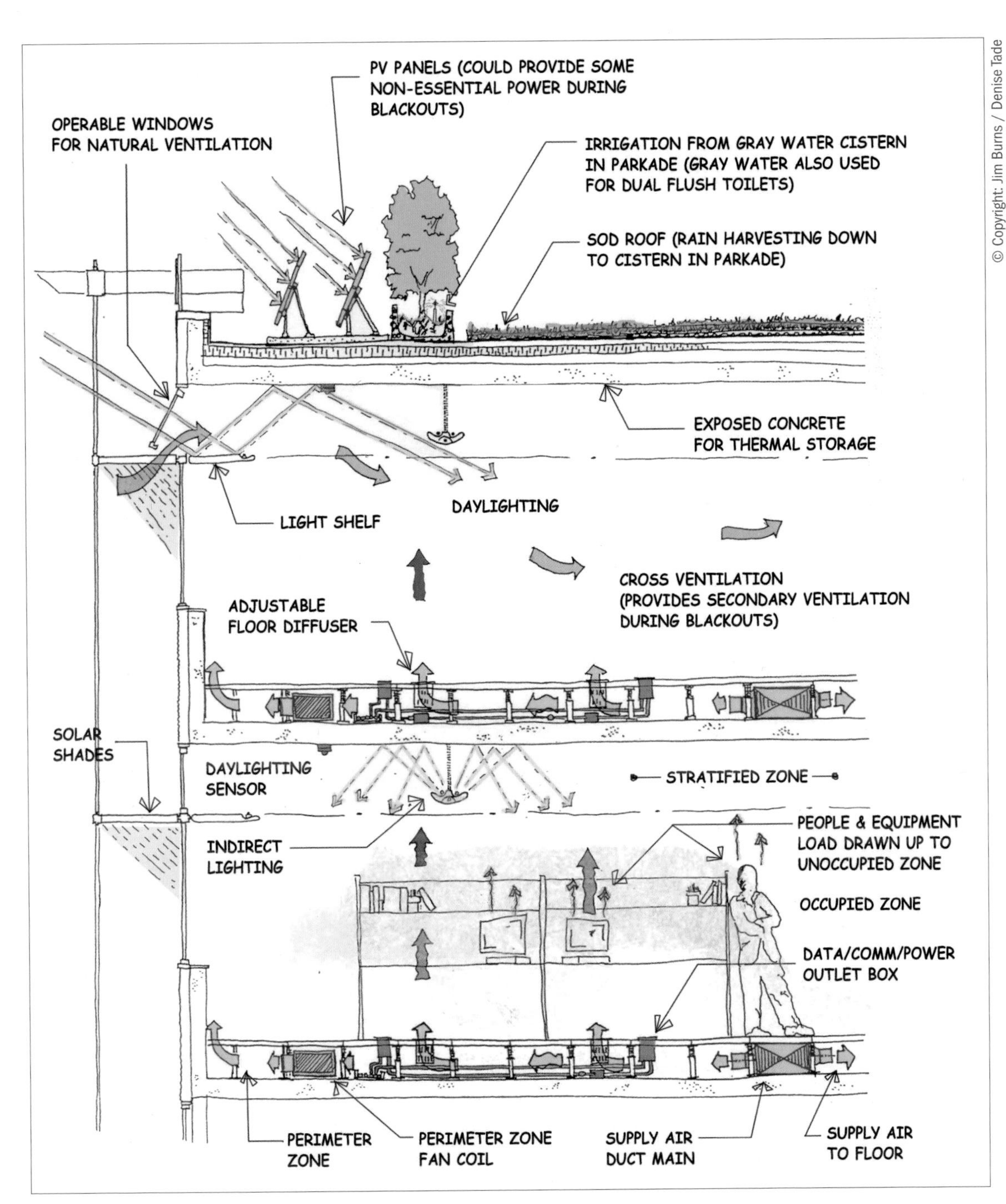

Far Left: The David and Lucile Packard Foundation Sustainability Matrix designed by KEEN and BNIM Architects. (2003).

Left: Making it fun and interesting: Over the last seventeen years, KEEN has created hundreds of concept sketches, or "cartoons," to help clients visualize an array of proposed energy, ventilation, lighting, heating and cooling schemes.

The work of an engineer is extremely important and, properly presented, can be very exciting – especially when the previous principles are followed and a great deal of integration and synergies between systems are occurring. The ecological engineer understands this dynamic, and seek to celebrate engineering – be it by exposing structural systems or air delivery systems or by finding other ways in which to convey the most important ideas and technologies. People should understand why their homes, offices and factories look and operate the way they do, because this knowledge exchange is essential in improving the environmental performance of projects.

At KEEN the task of visualizing systems is done on almost all projects. Visualization often takes the form of "cartoons," created by Jim Burns (KEEN's head of marketing) and his staff, that cleverly illustrate how each building is functioning at a level of complexity that nearly everyone can understand. When KEEN staff members can distill their own projects down to a basic snapshot, it has the benefit of providing a communication tool for their own design teams and their clients. As Burns says, "if I can understand it and draw it, then others will understand it as well." These cartoons are also incredibly fun, and convey engineering in a way that captivates and demystifies their work. KEEN took this process of visualizing information a step further in co-creating the Packard Matrix with BNIM Architects for the David and Lucile Packard Foundation. The Packard Matrix is a powerful tool which helps make visible an abundance of information related to achieving various levels of environmental performance. In this way the "result" of the engineering process is revealed – showing people how many households of energy are consumed and the amount of pollution that results from various levels of environmental performance. When people see the Packard Matrix or system cartoons produced by the firm, they gain a great deal of understanding about the importance of engineering and the need to understand and celebrate it.

The Methodologies

It is what we think we know already that often prevents us from learning.

– Claude Bernard

The last chapter explored the most important principles that guide the work of KEEN Engineering, making the firm an exemplary model for our concept of the "Ecological Engineer." Equally important to the idea of the ecological engineer are the methodologies that inform their engineering practice.

Like many things in life, there are multiple ways to engineer a project and principles like those outlined in the previous chapter could be achieved differently by different ecological engineers. This multi-faceted approach is certainly true for KEEN, where solutions are reached in a myriad of ways. However, what is always consistent on successful projects is a well thought-out methodology that informs the process.

This chapter highlights a few of the methodologies that inform the work of KEEN Engineering and provides the framework for how key decisions are made on projects of all types. To begin with, it is important to acknowledge that with KEEN, as with other ecological engineers, innovation and passion guides their work. When there is "purpose" behind a job, beyond just typical financial and personal motivations, people rise to the occasion and produce higher quality solutions not possible when people are simply going through the motions. KEEN's embracing of their collective responsibility to the natural world and clients' well-being pushes them forward to be smarter, leaner and better. This embracement is in itself a methodology. Individuals are attracted to KEEN because of their convictions and deeply-held beliefs, which engenders a culture of excitement and constant improvement.

As a company, KEEN works in two primary ways:

- As a consultant to an architectural firm providing a combination of mechanical, electrical or plumbing services as the Engineer of Record.
- As a consultant to other engineers or clients in a specialty role such as providing energy modeling, LEED® consulting or peer review while not serving as the Engineer of Record.

The following methodologies apply equally well to both of these consultant scenarios.

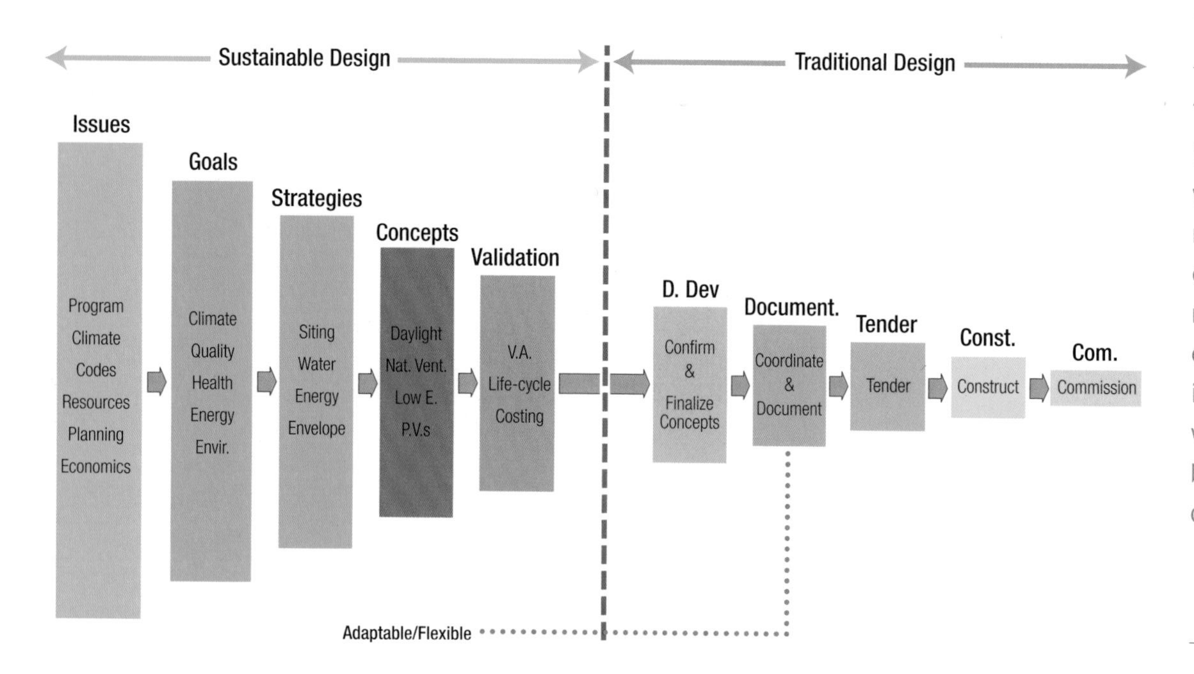

> The Green Decision-Making Matrix evolves and changes with each client. You've got to really understand the essence of each project, even if it's not in your discipline. You can actually get in there and influence that system and the way that decision is made by looking at the benefits of doing it in a different way.
>
> – Mark Mitchell, KEEN

KEEN Methodology One: Getting Involved

In this section reference will be made to the diagram shown above. This diagram represents a process that KEEN Engineering often uses in order to advance their principles on given projects. The diagram is divided into several individual components which describe a range of possibilities depending on how early KEEN can get involved in a project – the earlier the better, which is graphically highlighted with larger "bands" that represent increased possibilities. The farther along a project is, the more restricted are the opportunities. The methodology reminds KEEN to enter as early in the process as possible – and what to look for and attempt at each stage.

Issues – KEEN tries to get involved as early as possible on projects because they know they add tremendous value and can raise issues that other professionals or client groups might not consider. Most engineers are not brought into a project until much later in the process, after a project has been carefully defined, sites selected and the team assembled. KEEN recognizes that at the issues stage they can often provide the most value and help clients avoid very costly mistakes down the line. As KEEN's reputation has grown, they are increasingly brought in earlier as trusted advisors by many clients. It is at these early stages that the ecological engineer can most effectively challenge fundamental assumptions that affect every project - like climate, site selection, access to labor, and outside-the-box thinking about potential solutions.

The Ecological Engineer's lesson: *Do not think you can add value only after the building is designed. Strive to enter the project early and help define it. This early design stage is when environmental impact can be lowered significantly without spending a single dollar on construction.*

Goals – Getting involved in the upfront process when goals are being set is almost as good as getting involved when the project is just getting defined. Ecological engineers can help shape and describe what it means to produce a successful project. They can help determine what is achievable and stretch the comfort zone of their clients and colleagues to be more holistic and sensitive to environmental issues. By being involved early, other benefits are realized. Engineers can learn how best to serve their clients by listening to their stated values and by identifying opportunities to educate them when the opportunities present themselves. This opportunity identification cannot be done as effectively when values are described through a third party such as the Architect of Record. It is in the early phase of a project that LEED® is often mentioned, and the ecological engineer can help decide if it is appropriate as a tool and, if so, which LEED® program and which strategies should be pursued.

The Ecological Engineer's lesson: *Helping your client set goals for the project adds rigor and quality and can set the project in the right direction while providing useful feedback at the outset to inform your own design. Participate directly in goal setting if possible.*

Courtesy of KEEN Engineering

Whether leading it or as a participant, the design charrette has been invaluable to KEEN's understanding of building goals and alternatives in collaboration with the client, architect and other members of a project team.

Strategies – KEEN enjoys partaking in a collaborative, integrated process, which they understand makes for a more successful project in the long run. KEEN jumps at the opportunity to participate in design charrettes where clients, architects, engineers and other disciplines can get together and explore ideas and look for synergies and conflicts. In the words of Jennifer Sanguinetti, a KEEN project manager, "this is where we shine." The ecological engineer embraces the charrettes as the main engine for eco-innovation and integration that makes sustainability affordable and effective. KEEN looks to these opportunities to be a leader in the design process rather than a passive participant. When charrettes are not happening with enough frequency, KEEN tries to initiate such a meeting. When design meetings do occur, they work hard to always be included. Engineering should not happen after the building is designed but rather throughout the entire process.

The Ecological Engineer's lesson: *Look for ways to become an active participant in the design process by encouraging design charrettes. An interdisciplinary approach results in more integration and a better design.*

Concepts – As the design progresses, KEEN finds it important to present multiple options to the client and architect for the interior systems to be chosen for the building. By showing what is possible, rather than shoehorning a biased solution and leaving out alternatives, KEEN increases the potential for innovation by involving other disciplines in their own decision-making process. KEEN understands that it is not in their own best interest in the long run to "pessimize" certain options in order to railroad a design team into their favorite or comfortable solution. They believe in presenting options early so that if certain solutions play an important role in the form of the building then the design can still accommodate them. Presenting options in a format where the pros and the cons can be compared forces KEEN to stay open to possibilities and not draw conclusions too early. It also expands the mindset of the other members of the design team, and helps them appreciate the nuances of any particular engineering solution – be it mechanical, electrical or plumbing. Typically, alternatives are presented so that clients can see how each solution affects first costs, operating costs, and environmental impacts in terms of pollution, human health and aesthetics.

The Ecological Engineer's lesson: *Stay open to the possibility of design and do not immediately assume a solution. Clearly and fairly present as many feasible alternatives to the design team as possible, helping to educate all participants on the benefits and drawbacks of each solution. Better solutions often come from the dicussion.*

Validation – At multiple times throughout the design process, engineers at KEEN remind themselves to check back with the client in order to make certain that their needs are ultimately being met. Throughout the design process many changes, sometimes seemingly small ones, are often made. However, small changes can, at times, have profound effects on the satisfaction of the client with the engineering services for the simple reason that it was different than what was expected. Many engineers prefer to minimize face time and this approach can lead to significant problems and ill feelings after the building is occupied if the user groups do not understand or appreciate why certain decisions were made. A systematic process for checking with the client can be created to ensure a positive relationship.

The Ecological Engineer's lesson: *Do not make assumptions. At frequent intervals during the design process, find an appropriate opportunity to check with the client to ensure they understand and accept engineering solutions that evolve through the design sequence.*

KEEN Methodology Two: The Order of Operations Thinking

The most talented ecological engineers understand that being successful requires more than just specifying the right equipment or relying on certain engineering principles for a given project. True success results when engineers do things in the right order. Like a mathematical equation, if the right order of operations is not followed the results are usually less than optimized performance at greater cost – not the best solution. The Order of Operations Thinking[1] is a powerful methodology for ensuring the best fit between performance and cost. As Blair McCarry points out "you have got to do things in the right order or you get the wrong result – it's just common sense." A summary of the methodology is as follows:

Reduce Loads - Order of Operations Thinking always starts with trying to reduce the loads in a given building. For the mechanical engineer this thinking means finding ways to reduce heating and cooling loads, because the best way to save energy is to not have to provide as much heating and cooling in the first place. For the electrical engineer, reducing loads can refer to carefully understanding how much light is really needed in various spaces rather than assuming a blanket amount of light is needed everywhere. For the plumbing engineer, it means investigating options that reduce the amount of water required within the facility. Often times engineers skip over this step because it gets into areas traditionally "outside their control." Architects typically control a building's envelope conditions which play a huge role in how much energy a building will use to maintain interior temperatures. Users have expectations for interior temperatures even though they do not understand that comfort is not just tied to interior temperature[2]. An opportunity for leadership is available as engineers can work with other disciplines and the client directly to find ways to reduce loads, primarily by helping them early in the design stage to understand the implications of their basic assumptions and design moves. Reducing loads is almost always the most effective and cost – efficient way to save money and reduce energy or water use.

The Ecological Engineer's lesson: *Always start with load reduction strategies, even if that means challenging the client's and architect's assumptions. Load reduction is the most powerful way to reduce the environmental footprint of a project.*

Use Free Energy – Once an engineer has helped reduce loads in a building, an opportunity exists to find "free" sources of energy and water to meet some or all of the remaining demand. As with the first step, this step is often overlooked by traditional engineering practice that goes immediately to the "solution step," bypassing many opportunities that can result in superior comfort and performance. Daylight, if handled properly, can result in superior visual acuity at greatly reduced energy costs provided that the electrical engineer designs control systems and lighting to respond to available daylight. Natural ventilation, passive solar heating and rainwater collection are just a few more of the low-tech solutions that are often overlooked. Low-tech solutions will be discussed in more detail in the next chapter.

The Ecological Engineer's lesson: *Embrace low-tech solutions at every opportunity and do not assume that mechanical or high-tech solutions are always superior. Always start with load reduction strategies first.*

Use Efficient Technologies – The last step in the methodology (and the first by most "efficiency engineers"[3]) is the selection of efficient and appropriately-sized equipment. The most efficient solutions often carry a higher premium than conventional equipment. However, if the first two steps are done in the right order, the final solution can often be much less expensive overall as everything gets smaller. The ecological engineer does not just assume that something is too expensive at the outset, but instead looks to find ways to reduce the demand to the point where it is cost effective or even cheaper. A few high-tech solutions to problems are explored in the final chapter of this section.

The Ecological Engineer's lesson: *Do not optimize a technology without thinking about the whole building as a system. Always start with load reduction strategies and then find ways to use natural or "free" solutions. If done as the third step, and properly sized, the most efficient high-tech solution is likely to be affordable.*

Using LEED® as a Methodology

From almost the outset of the creation of the US Green Building Council, KEEN has been a big supporter of the LEED® program, even though it was not until 2004 that Canada adopted its own version of LEED®. KEEN, like many of the most astute ecological engineers, realized that LEED®, despite having some improvements to work out, was the most powerful tool ever to come along in the building industry in terms of the potential for market transformation toward more responsible building practices. KEEN embraced it at all levels. KEEN staff members started by first making themselves experts and creating an in-house program that has resulted in almost everyone in the firm becoming a LEED® accredited professional. They then sought leadership positions by volunteering for committees. A culmination of these leadership positions was realized when Kevin Hydes became the board president for the USGBC in 2005.

For many organizations, LEED® is something that is used only because of market pressures. But KEEN uses it as a driving methodology within the company, even when clients are not demanding certification. KEEN realized quickly that LEED® provided rigor and tangible benchmarks within a holistic framework – making sure that all aspects of sustainability are considered and goals carried through the life of a project. The LEED® methodology is a required checklist on virtually all new projects, and it is even useful for reminding other team members of commitments and goals made at the beginning of the project. In the words of Paul Anseeuw, KEEN Seattle's leader, "It helps put things back on the table."

KEEN has been involved with more LEED® certified projects at various levels than any other engineering firm in North America, but more important is how LEED® has shaped the mindsets of the individuals within the organization. It has become a series of questions to ask – and a continual reminder on what issues are important. Support comes strongly from the top and so LEED® is taken very, very seriously. When projects do pursue certification, it is usually the KEEN Concepts team that does the documentation or serves as a resource for LEED®

documentation. As a sub-group of the company, team members enjoy the work immensely, and find the evolution of the LEED® tools compelling.

The Ecological Engineer's lesson: *Make LEED® an important part of how your firm does business. Become an expert on its usefulness for setting goals, educating clients, raising performance and achieving excellence. Do not wait until you have to use LEED® to become an expert. Start today.*

Cartoon as Methodology

In the last chapter, KEEN's cartoons were discussed as an important component of their principle of Interjecting Fun and Visualization into the Practice. At a primary level, they help to make the invisible visible and help to make sure that concepts can be properly explained to clients and architects. From a methodological standpoint, the cartooning process goes deeper into helping KEEN with their work and provides a useful lesson for any aspiring ecological engineer. The ability to draw is a powerful tool, not only for conveying information externally to non-experts, but internally as a clear communication tool that helps to make sure that the original design concept is not lost as a project moves through various stages to completion. Even if different individuals are assigned to a project, for whatever reason, a good cartoon or "engineering parti[4]," to use the lingo in the architecture profession, can be used to verify intent. Many engineers are afraid to draw and this fear limits their potential. A drawing does not always have to be pretty to be useful. It forces engineers to really think about how things are working. Jennifer Sanguinetti makes an interesting point: "Creating diagrams of airflow requires that you really understand how air moves and that you are not just 'training' arrows to do what you want them to do." Creating diagrams forces a greater understanding of interactions between architecture and internal and external conditions.

The Ecological Engineer's lesson: *Draw your design. Find ways to make designs and systems visible as a tool towards greater understanding for your team and for your clients.*

Computer Modeling and Analysis as Methodology

In the last decade, computer simulation tools have increased in sophistication and ease of use. More and more engineers have begun using various programs to help them in their engineering process. Often, the model is used to verify or predict energy usage or tell the "answer" at the end of a design process. More astute engineers are realizing that modeling tools are much more useful as early design tools than as end-of-project "predictors" of energy use. Used correctly, tools such as DOE-2, E-Quest and Energy-10 can become a powerful method for understanding the complex interactions of a building's systems and, through careful parametric analysis, can help a team make truly informed decisions about the relative merits of various strategies.

Modeling should never be used just to justify a scheme in a biased way unless that scheme truly is the best one when analyzed in a non-biased and scientific way. Good schematic level energy models can be used to direct design, and help the team understand impacts very early on. KEEN understood the importance of energy modeling, and many individuals in the firm have become experts in modeling building performance. They try to use these models as a design tool on all projects they undertake and use them to compare the pros and cons of various strategies and give educated feedback to the architectural team on design decisions.

More recently, KEEN embraced thermal analysis software in order to more accurately assess human comfort issues that cannot be done in traditional energy software and to better understand airflow, especially in passive ventilation schemes that traditional software is completely incapable of modeling. Other tools in their toolkit, including Transys and Ecotect, are used to understand the feedback between loads and systems and are also useful for understanding thermal mass and radiant systems. Ecotect helps team members to perform shading and comfort analysis. The modeling methodology is consciously used as a tool and its limitations are clearly understood. According to Hydes, "Modeling

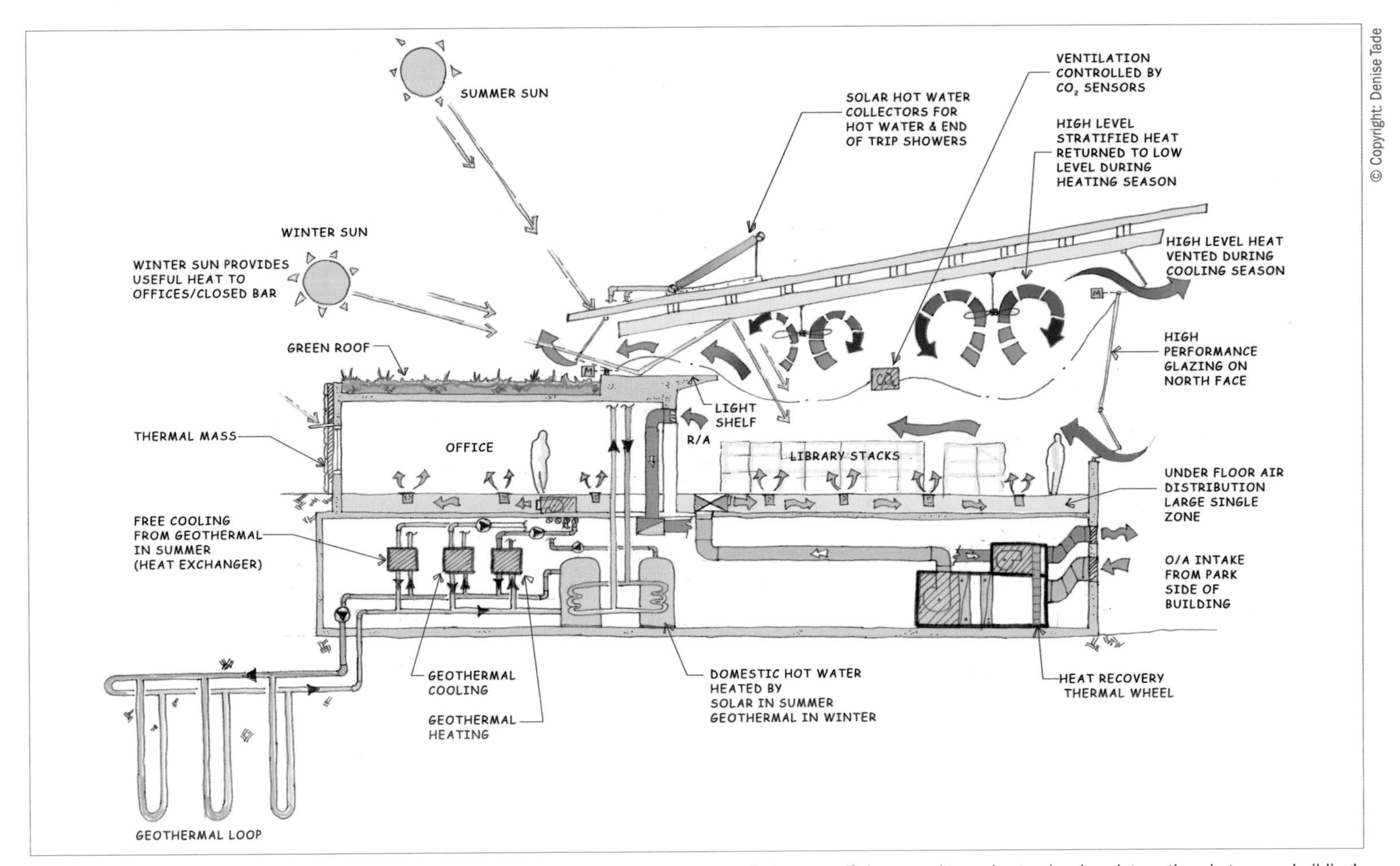

Diagramming all mechanical and electrical schemes – a methodology KEEN Engineering has relied on to verify intent and to understand various interactions between a building's architecture and site conditions (Whistler, BC Library by Hughes Condon Marler Architects).

can never be used as a crutch, and doesn't replace knowledge of the basics."

The Ecological Engineer's lesson: *Use modeling software as a design tool throughout the process, but always be sure to use the proper tool for the job. Be aware of its strengths and limitations and apply it for a more scientific understanding of building performance.*

Quality Control and the Post-Design Methodology

Any good engineer understands that quality control is critical to enduring success. A lot happens through the construction process, and informed engineers will do what they can to understand what worked and what did not work so that mistakes can be fixed and lessons learned for the next job. The same quality control focus is true for the ecological engineer who understands that malfunctioning systems or unhappy occupants always result in greater environmental

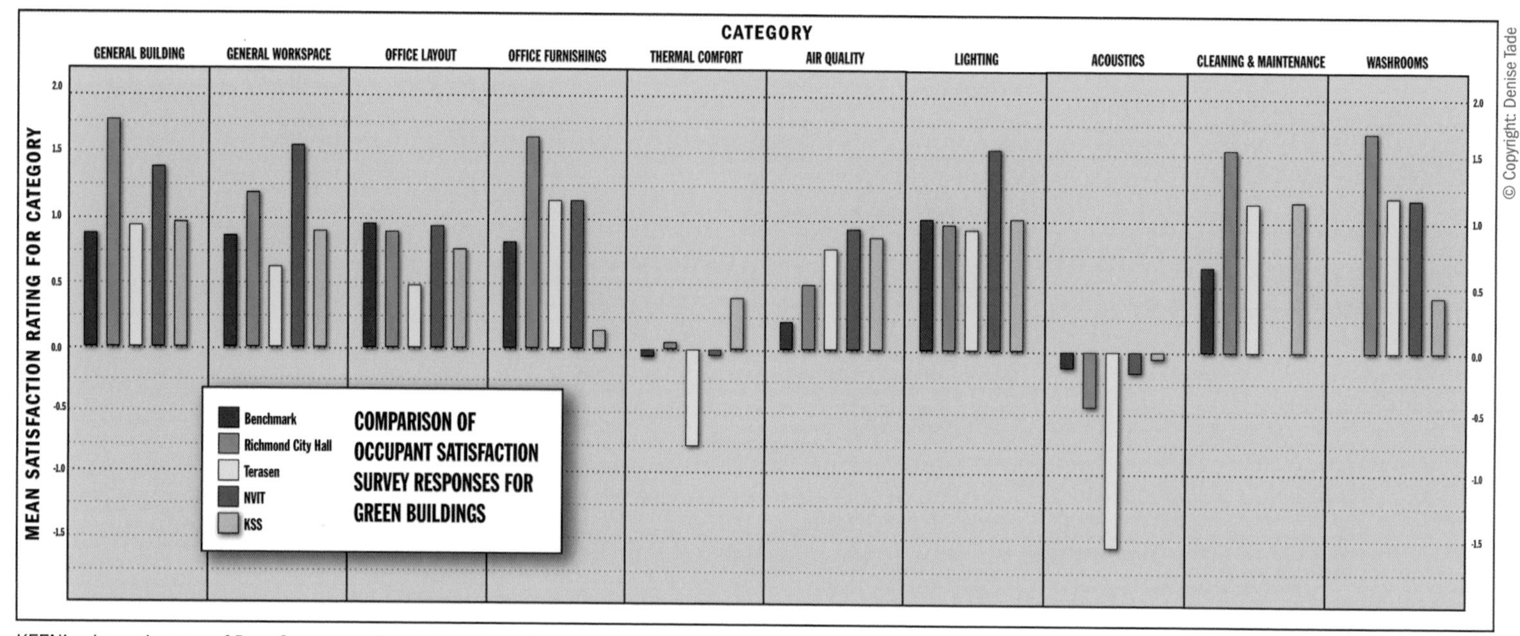

KEEN's pioneering use of Post-Occupancy Evaluations (POEs) is enabling its engineers to better gauge client satisfaction and benchmark design practices and technologies.

impact and reduced productivity.

A methodology of quality control is important at KEEN Engineering and various types of Post Occupancy Evaluations (POEs) are performed on critical projects as a tool for checking with clients to see how satisfied they are with their systems. This relationship-building process also provides them with valuable opportunities to check earlier assumptions and modeling calculations against reality – invaluable information for making the next project better. Like any engineering firm can attest, finding funding for a full POE is often difficult because clients do not always understand its value, and so KEEN does what it can, when it can, including self funding selected POE analysis when feedback is critical. KEEN sends one-page memos to their entire staff that provides critical lessons learned from built projects. These memos are kept short so that the information is read and digested.

On some projects, like the IslandWood Project, KEEN has the opportunity to create a training manual that helps their clients properly operate and maintain their systems. At other times they get an opportunity to provide commissioning services on their own projects and on other engineers' projects where there is a great opportunity to learn. All of these analyses and staff memos are efforts aimed at learning how to provide superior service while lowering environmental impact.

The Ecological Engineer's lesson: *Meeting stringent levels of environmental performance requires a methodology of follow-through past the construction of a project. Be sure to conduct a review after the project is completed. Find innovative ways to learn from past projects and provide better service to your clients.*

Low-Tech Solutions

Natural Ventilation, Thermal Mass and Façade Integration

The vital necessity of perfect ventilation, an even temperature, the maximum of pure air and the minimum of poisonous gases and dust, is too well known to need argument.

– Ford *Factory Facts* booklet, 1912

It is not uncommon to think of today's mechanical or electrical engineer as someone who uses technology – and lots of it – to design building systems and troubleshoot HVAC problems. Conventional wisdom holds that an engineer's solutions are almost always "high-tech" and mechanically intensive, using invention to improve upon natural processes.

However, the ecological engineer is someone who first looks at natural solutions to solve challenges rather than immediately assuming that a mechanical or human-based invention is always superior. The ecological engineer embraces innovative "low-tech" or natural solutions, continuously pursuing reduced operating loads – and often better solutions.

An increasing number of studies support the positive effect of sustainably-designed workspaces on performance and productivity, particularly those emphasizing natural air and light. Judith Heerwagen, an environmental psychologist in Seattle and senior scientist at the Pacific Northwest National Laboratory, has written that recent field studies on sustainable design "show the thermal environment (especially temperature, humidity level and ventilation) influences task performance, air quality and acoustics"[1]. Among the documented findings are that:

- personal control over ambient conditions is associated with enhanced work performance;
- increased incidence of illness associated with poor indoor air quality reduces performance on tasks requiring alertness and attention;
- and improved connection to the natural environment through views and outdoor amenities reduces stress and improves overall psychological and emotional functioning.

Certainly, KEEN Engineering has earned its reputation, in part, through an embrace of low-tech ideas that utilize the wind, the sun and specifics of microclimate and climate to create super-efficient buildings as well as comfort for clients. This chapter explores three of KEEN's innovative practices as they relate to low-tech solutions:

- Natural Ventilation
- Thermal Mass
- Façade Integration

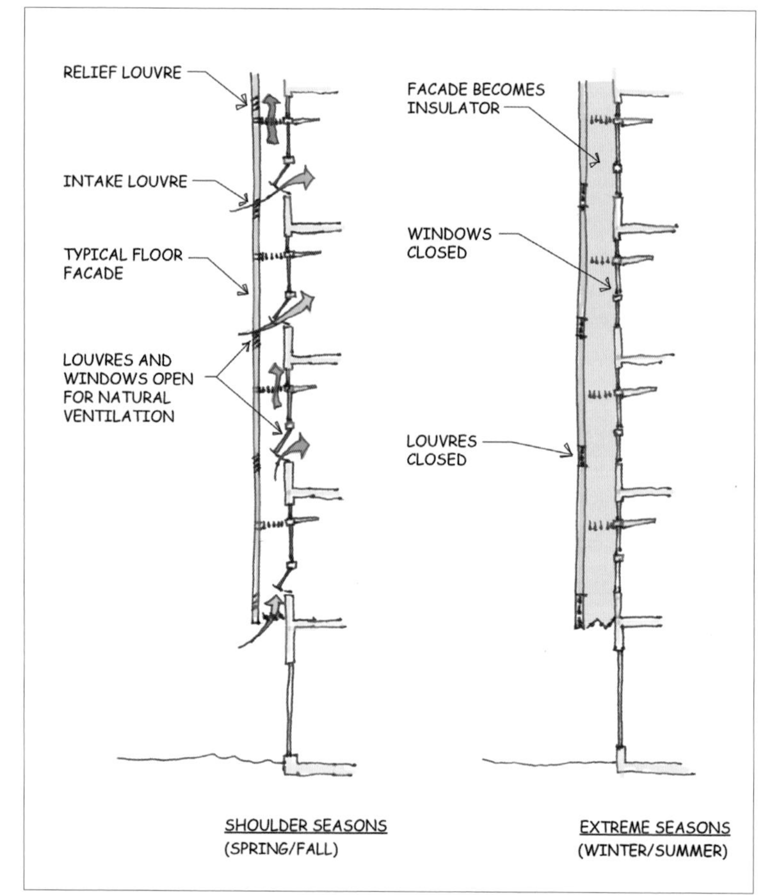

A KEEN "cartoon" demonstrating how a double envelope façade works in each season.

Part One

Breath of Fresh Air

Before there was air conditioning and before commercial buildings used "economizers" to draw in fresh air as outdoor conditions permitted, there was natural ventilation. Kevin Hydes particularly likes to reference the famous General Motors Building in Detroit, circa 1924. This pre-refrigeration, pre-air-conditioned structure, like others of its time was designed to open up and permit airflow so that people could be cooled by breezes and gain constant access to fresh air.

"When we talk of 'going back to the future' in terms of design, we just need to look back at all the pre-1930s buildings designed in Kansas City, Chicago, Detroit, or any of the great industrial cities," said Hydes. "All of the great architecture that came out in every one of those buildings serves as a model for natural ventilation."

> Even at the concept stage, we started thinking of the C.K. Choi Building like a church. We saw these atriums as our natural ventilation pools. So we started doing some simple calculations: how many people were in the building and how many cubic feet do they already have to breathe... The benefit of all the oxygen that's already in the big volume can contribute towards the ventilation and effectiveness of a space.
>
> – Kevin Hydes, KEEN

With the rise of air conditioning, it became possible to provide cool interior temperatures regardless of how hot it was outside or how much heat was generated inside. It also became possible to tightly regulate temperature and create what was thought to be the "ideal" conditions for human comfort, typically about 72°F. With air-conditioning came humidity control as well–another significant comfort advantage. Relief was provided to places that had long suffered from incredibly high temperature/humidity combinations and, since energy was cheap, the change was readily welcomed. Because air conditioning "normalized" outside and inside conditions, it became possible to design buildings with almost any form and size. No longer were narrow footprints and operable windows necessary to create comfort. As a consequence, buildings became bigger and wider, with a greatly reduced perimeter-to-area ratio.

Curiously, even in locations where natural ventilation was incredibly effective, it was abandoned in favor of expensive and complex mechanical climate-control systems. Fads, status and the perception of superior comfort provided a triple threat to the art of natural ventilation to the extent that it virtually disappeared in commercial buildings in the U.S. This near-complete reliance on purely mechanical means by the AEC industry during the 1970s, 1980s and 1990s was partly justified because, as expectations for comfort increased, so did complaints and even litigation. It became necessary for design professionals to create systems that were tightly controllable in order to ensure sufficient cooling capacity and reduce complaints. At least that was the argument. Using copious amounts of energy was deemed an acceptable trade-off for capacity. Through these types of modern design "choices", natural ventilation has, in many ways, become something of a lost art – a concept only recently revisited as engineering firms like KEEN show that it is possible to create highly functional, naturally-ventilated buildings within the construct of modern expectations of comfort.

A Closer Look at Natural Ventilation

Successful natural ventilation requires a deeper understanding of the subtleties of this design technique through study of existing naturally-ventilated structures and exploration of tools such as computational fluid dynamic (CFD) modeling that help make invisible airflow visible. In essence, natural ventilation design should be based on several key principles and strategies – beginning with appropriate climate.

Climate is a critical, but often misunderstood, aspect of natural ventilation in buildings. Put simply, natural ventilation works best when it is cooler outside than inside. This critical climatic aspect calls for local conditions where the average daily temperature does not exceed 70°F during summer months and the average high is 80°F or less with reasonably low humidity levels. When temperatures are hotter than this, natural ventilation will not work. In some climates daytime temperature can be above the accepted range, but they must then drop significantly enough at night to cool a building with natural ventilation.

In most locations across the U.S. and Canada, at least some part of the cooling season falls within an appropriate temperature and humidity range for natural ventilation. The Pacific Northwest, where KEEN's head office is located, is ideal for exploring these strategies; in fact, careful design in that mild climate means that buildings can be completely naturally ventilated if properly designed. Mountain and desert climates are also appropriate for natural ventilation, since they tend to have large diurnal (day to night) temperature swings. Even on the hottest days, buildings can be completely purged at night due to the significant drop in temperature. KEEN is also seeing interest in cold climate natural ventilation along the eastern seaboard, further inland in upstate New York, Vermont and New Hampshire, and in Canada's eastern provinces.

Predominately hot, humid climates pose the greatest challenge, particularly those of the central states and southeastern U.S. Yet, even in these regions where temperatures are low enough during swing seasons, a well-designed structure can be ventilated at night and then "float" through the day or be open all day if temperatures permit. Hybrid designs, employing both natural ventilation and conventional air conditioning techniques, also show promise but offer less initial savings.

Challenges/Opportunities

It is ironic, then, that many buildings along North America's west coast still rely so heavily on mechanically-intensive systems. Natural ventilation could be used effectively anywhere within ten miles of

Case Study

The Liu Centre

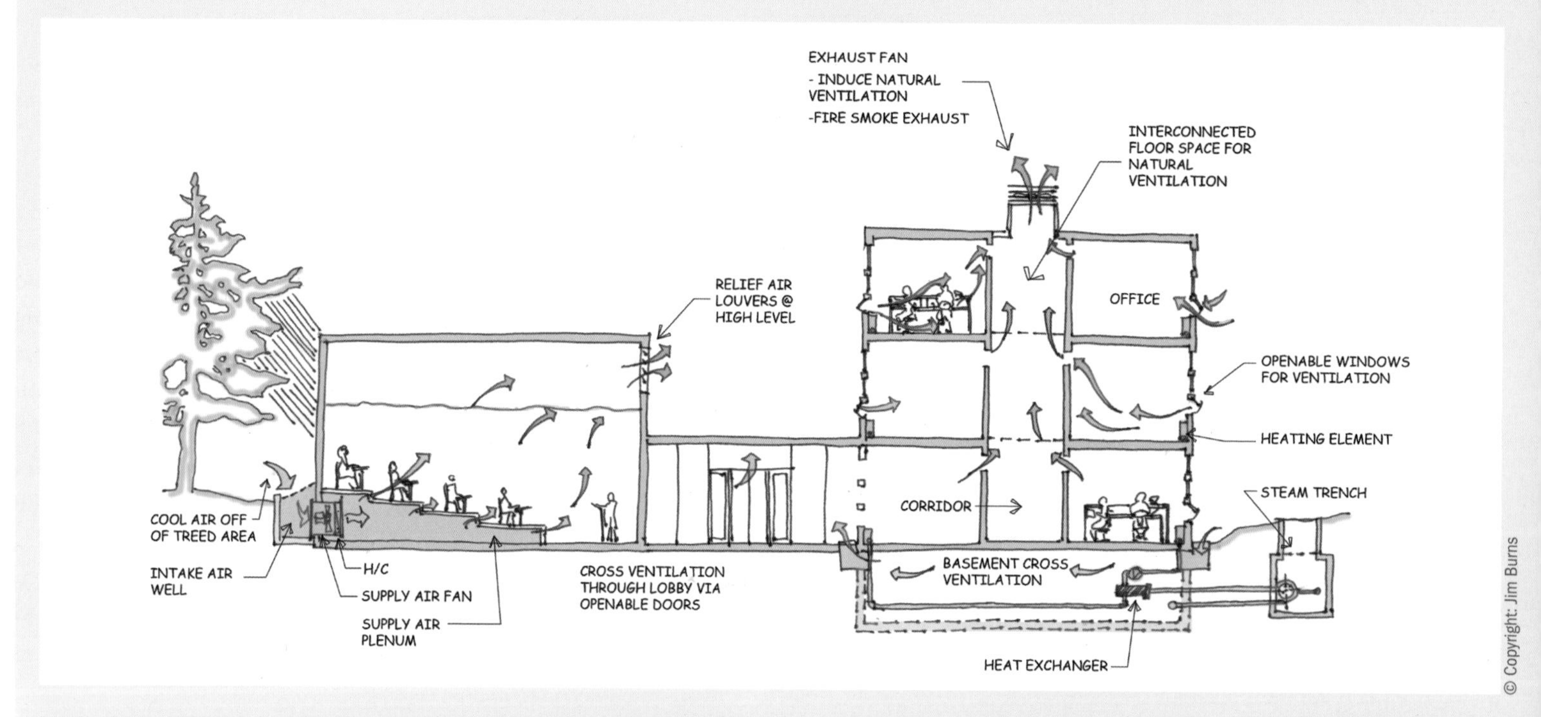

In the late 1990s, the University of British Columbia constructed the Liu Centre for Asian Studies on a site adjacent to the C.K. Choi building in Vancouver. Stantec Architecture, in collaboration with Arthur Erikson was selected as the architect, with KEEN as the engineer of record. In many ways, Liu was considered the "child" of Choi, so that the University and project team fully expected to apply many of the design lessons learned on that earlier project. Ultimately, the completed Liu building did achieve very similar results in terms of materials re-use, while meeting aggressive targets for reduced energy and water consumption.

The Liu Centre was conceived on a similar scale to the Choi Building, with similar occupancy and usage expected. Nestled in a giant stand of trees, the building site created a suitable microclimate for natural ventilation, which became a critical component of the design. KEEN had recently acquired an important tool to predict ventilation performance in buildings. This new simulation software allowed its engineers to create a number of "what if" scenarios based on building configuration, projected airflow and location within the forest that, in turn, helped to inform the architect and the final architecture.

As designed, the Liu Centre includes considerable thermal mass and permits airflow into the interior of the building, also narrow in depth. Perimeter windows are fitted with "trickle" ventilators (small 2" high openings integrated into the length of the window units) that can be opened and closed to permit air. Hopper windows are also

© Copyright: gerry kopelow / photographics inc.

included for larger intakes as needed and controlled by the user. As air flows into perimeter spaces it is vented into interior hallways through transom openings located above doorways. Daylight is also distributed to the interior in the same manner. Once air enters the heart of the building, hot air rises up into exhaust chimneys that utilize the stack effect.

Low-Tech Lesson I: *In the Liu Centre the building's cooling system is controlled not by high-tech "gadgets" or fan power, but by common sense – the ultimate low-tech solution. Users are instructed on how to use the building – and the result is an effective natural ventilation solution. Do not always assume that high-tech controls are always superior.*

Low-Tech Lesson II: *The specific site chosen for the Liu Centre, tucked further into the native northwestern forest, played an important role in reducing cooling loads. Engineers should seek to understand microclimate conditions to fine-tune building design.*

the ocean, from British Columbia, Washington and Oregon down to southern California. Hydes noted:

> That's 40 million people living in that population strip who could work in naturally-ventilated buildings, offices and schools. It's completely do-able. So we're part of a larger effort, really, that's helping to convince entire communities on the West Coast that if natural ventilation works in your house, it can work in your office. That's where the design of these buildings is comparable to just designing big houses.

Building occupants who have experienced natural ventilation, whether in historic or modern structures, cite improved air quality, productivity and overall comfort as primary benefits. Recent satisfaction surveys are also demonstrating the importance of having personal control over a workspace – simply the ability to open and close windows. Equally significant for building owners is that natural ventilation and cooling strategies can greatly reduce energy costs – as much as 10 percent to 25 percent – versus conventional systems. These savings result from lower electricity consumption, less ductwork and reduced floor-to-floor heights in building construction.

Still, there are obstacles – real and perceived – to the use of natural ventilation in buildings that rely on operable windows or vents. These obstacles include concerns about managing noise, the quality of outdoor air, rain penetration, intruder prevention, and the potential for wasted energy when occupants forget to close windows or ignore established schedules. Smoke control in case of fire is another potential issue.

In addition, mechanical engineers traditionally schooled and experienced in designing systems for "closed" buildings may remain skeptical. It has been their professional responsibility to meet thermal comfort standards such as ASHRAE Standard 55 or ISO7730 as commonly demonstrated through controlled mechanical ventilation. In some cases, these practices originated with the oil crises of the 1970s, when codes called for under-ventilating buildings to save energy and later revised in the 1980s. As some journals are pointing out, though, many current local and national standards define indoor thermal comfort so narrowly – even for mild climates – that it is sometimes difficult to address what is thermally acceptable for buildings without mechanical systems. Today, the question for KEEN and other ecological engineers – in terms of comfort, occupant satisfaction and energy savings – increasingly becomes: "Can we trust the codes?"

Best Practices

Undoubtedly, natural ventilation is easier to design and implement in some building types than in others, due to the range and size of internal cooling loads; for example, computers, monitors and photocopiers, as well as electric lighting and people, generate internal cooling loads. Structures with very high plug loads typically fall outside of what is possible with natural ventilation. These structures include buildings such as theaters and auditoriums that experience large concentrations of people, as well as large, equipment-intensive spaces such as factories and data centers. Very large buildings with deep floor plates and/or those subdivided into multiple "cells" are also difficult, if not impossible, to naturally ventilate. Nevertheless, most building types including homes, offices, schools and retail stores are appropriate candidates for this design strategy. In North America, most buildings constructed before 1950 were designed for natural ventilation, so natural ventilation often makes sense to consider for renovations to return the building to its original state. New building types with less stringent climate-control requirements are also strong candidates.

Natural ventilation design falls into two basic concepts: stack ventilation (buoyancy) and cross ventilation (wind). The buoyancy, or "stack effect," system relies on gravitational forces and the buoyancy of warm air to draw air up and out of the space. The wind, or cross ventilation system allows supply air to flow from one space to another due to pressure differences between the windward (positive) and leeward (negative) sides of a building. Stack ventilation utilizes temperature differences and building forms to direct airflows through corridor connections to atriums or open central stairwells. A "solar chimney" may be added to the exhaust to enhance the stack effect. In contrast, cross ventilation is primarily wind driven using operable windows, shutters and louvers – openings oriented toward prevailing winds to maximize airflow through the building. Airflow rate depends on the size of these inlets and outlets, the height difference between them, outdoor (and indoor) temperatures, and other variables.

> Simplicity and elegant frugality are natural partners. Using less material means there is less to go wrong, less work involved, less cost, and better performance.
>
> *– Natural Capitalism: Creating the Next Industrial Revolution*

Green designers like KEEN believe that natural ventilation is most effective at maintaining comfort when occupants have some control of their individual spaces in combination with daylighting and cooling-load reduction strategies such as night flushing, shading, high-performance glazing, and sufficient thermal mass to dampen temperature swings. The performance of a naturally-ventilated building is ideally simulated and measured by tracking ventilation flow rates, indoor temperatures, CO_2 concentrations, and heating and cooling loads.

Based on its growing list of projects, KEEN Engineering also knows that natural ventilation can effectively co-exist with mechanical ventilation, mechanical cooling and heat recovery in low- to mid-rise commercial buildings. These mixed-mode, or "hybrid", ventilation systems offer the possibility of energy savings and improved comfort in a greater variety of buildings and climates. Mix-mode spaces may be designed to switch as needed from natural to mechanical ventilation on a daily or seasonal basis.

Design Considerations

"First engineering" principles are fundamental to natural ventilation: warm air rises, air moves from lower to higher pressure, hot to cold, wet to dry, and so on – always attempting to achieve a balance.

Case Study

Mithun Architect's Offices

© Copyright: Robert Pisano

When Seattle's Mithun Architects needed a new office, they creatively chose a pier to provide a dramatic setting for their designs – and they hired KEEN Engineering to provide innovative solutions for their MP needs. The existing pier, built with a wooden frame and a narrow floor plate, was historic in nature. Not much thermal mass was present in the building and yet, as KEEN realized, it was sitting on top of the world's best thermal mass – the ocean. The microclimate provided by a cool water surface, radiated to the underside of the structure, and breezes across the water made natural ventilation an even smarter choice. Further assisting a natural ventilation scheme was the original design of the pier: its barn-like form and high clerestory casement windows could be used to exhaust air. Indeed, as the design progressed, these windows were fitted with a gear mechanism so that people could operate vents at the top.

A key component of the natural ventilation strategy for the Mithun offices involved the recommended procedure for opening windows – because there literally was a right way and a wrong way with this design. Because of the pier's location near the Alaskan Way, a giant elevated roadway filled with noise and air pollution, if the wrong casements were opened first (or just one opened the wrong way) then the noise and the airflow from the highway would be redirected into the building. By opening the opposite casement, acoustic problems were greatly minimized and air coming off the water could be directed into the interior.

At the start of the project, KEEN used thermal analysis software to model airflow in the structure. And by the time building occupants were ready to move in, they were given clever "O&M" (operation and management) cards that described proper operating instructions for their new, fresh air space.

Low-Tech Lesson III: *Passive solutions can sometimes have unintended consequences, and they must be considered with as much care as any conventional system. Increased acoustic challenges and sometimes air quality problems can arise with natural ventilation. These challenges must be carefully understood so that the design solution is not compromised by other factors.*

Low-Tech Lesson IV: *Historic buildings were, naturally, designed for passive, low-tech solutions. In a retrofit situation, first analyze how the building was originally designed to work. Too often, successive retrofits have diminished the building's effectiveness for daylight, as well as for passive heating and cooling. By uncovering what was original and supplementing it carefully, a successful low-energy building is possible.*

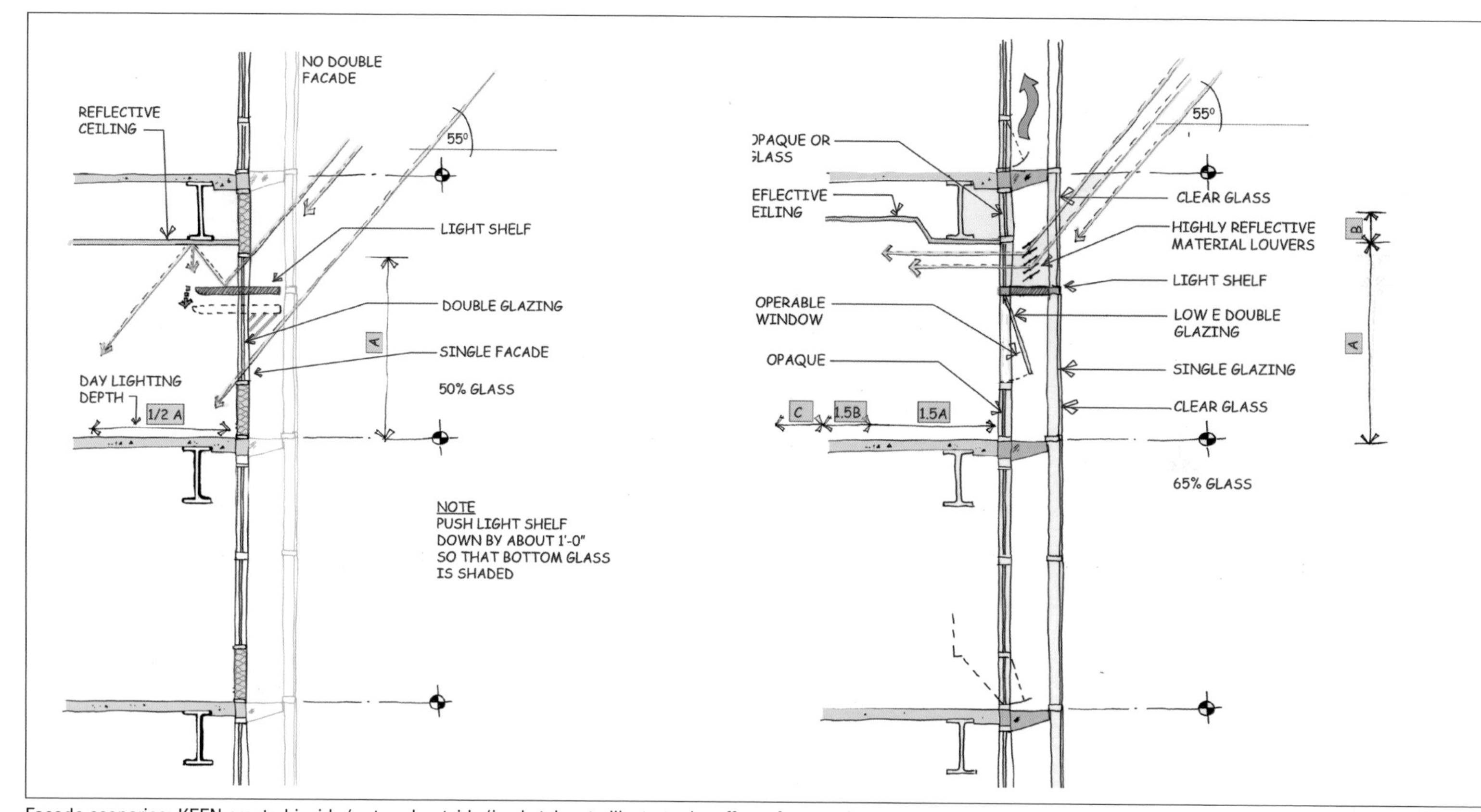

Façade scenarios: KEEN created inside/out and outside/in sketches to illustrate the effect of external shading, glazing, insulation and other variables on building design.

Proper design really begins with close collaboration between the engineer and project architect to determine various upfront options. Assuming an appropriate building type and climate, the first key is to consider orientation to take advantage of prevailing winds and to locate building inlets and outlets for optimal cross ventilation. The second key is building shape, enabling the structure to automatically ventilate itself, using buoyancy and gravity, under the right conditions. And the third key is environment – although, with adequately-designed filtration systems, even buildings in urban areas with high outdoor pollution should still be considered.

In terms of specific design elements, naturally-ventilated buildings should be sufficiently narrow – with floor plates forty feet wide or less – to allow for cross ventilation or tall enough to allow for the stack effect to drive airflow. Naturally-ventilated buildings typically require thermal mass to regulate interior temperature fluctuations. Inlets and outlets need to be properly sized and located, and the building's external heat gain should be carefully controlled through proper sun shading, glazing choices, insulation values and roof color. Buildings that are short, wide and built with low mass will negate cooling from natural ventilation.

Another important consideration in design is what KEEN refers to as risk sharing: educating building owners and users about natural

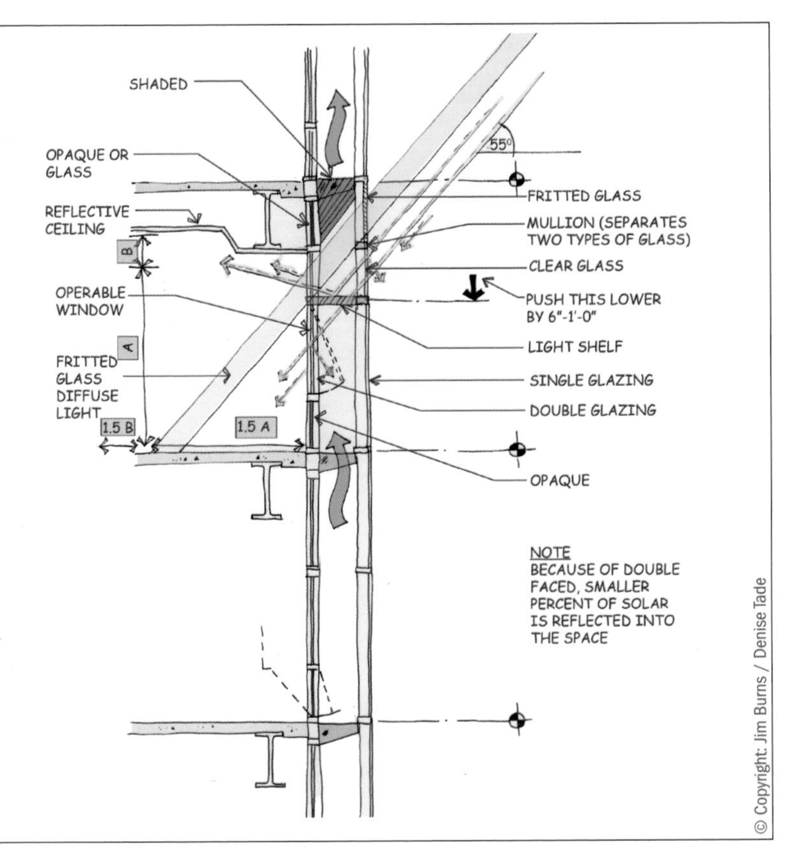

ventilation to broaden their expectations of indoor comfort, particularly with systems that differ from standard practice. KEEN believes this process is important in gaining buy-in and an understanding that occupants may sometimes experience (as a hypothetical example) 80°F temperatures over the course of a few hours each year in exchange for fresh air and personal control with operable windows.

Support is gaining, too, for broader definitions of comfort when using a technique such as natural ventilation in new or renovated buildings. In *Greening Federal Facilities*, published by the U.S. Department of Energy, the agency stated that "performance requirements should not include strict limits on acceptable indoor temperature and humidity conditions...because extreme weather conditions are difficult to predict." Furthermore, the agency said, "The more broadly these conditions are defined, and the larger the acceptable amount of time out of compliance, the greater the possibilities for reducing mechanical system size and usage."[2]

Further Natural Ventilation Resources

"A Standard for Natural Ventilation", Gail Schiller Brager, Ph.D., and Richard de Dear, Ph.D., *ASHRAE Journal*, October 2000.

Climate Considerations in Building and Urban Design (1998), Baruch Givoni, published by John Wiley & Sons, New York, NY.

"Natural Ventilation and Collaborative Design", Jim Jones, Ph.D. and Aaron W. West, *ASHRAE Journal*, November 2001.

Natural Ventilation in Buildings (1998), Francis Allard, editor, published by James and James, London.

Natural Ventilation in Non-Domestic Buildings (2005), from CIBSE, (the U.K.'s Chartered Institution of Building Services Engineers), is a major revision of their natural ventilation applications manual first published in 1997.

Sun, Wind, and Light: Architectural Design Strategies (1985), G.Z. Brown and Virginia Cartwright, published by John Wiley & Sons, New York, NY.

Part Two

In That Quiet Earth

Globally, the earth is the perfect thermal mass – affecting air temperatures near its surface with both diurnal and seasonal time lags according to the storage capacity of surface materials (rock, soil and water) and the re-radiation of stored energy. Long before modern mechanical systems and forced-air heating and cooling were ever conceived, stone, masonry and brick were the construction materials of choice. Traditionally, these materials slowed the heat transfer through walls and gave buildings the necessary mass to radiate heat for warmth in winter and absorb heat to provide cool temperatures in summer.

Thermal mass functions as a thermal regulator within buildings, helping to dampen temperature fluctuations by absorbing and releasing heat, depending on ambient air temperatures. It can also play an important role in comfort as well, as our bodies "sense" the radiant temperatures of the materials around them. If exposed to a cool mass, indoor air temperature set points can be raised in summer, thus reducing energy use. Conversely, winter temperatures can be much lower if occupants are exposed to enough warm thermal mass: the basic principle behind passive solar buildings. Using the sun to heat thermal mass is a perfect low-tech fit, synchronizing free energy and the thermal capacitance of materials. To be effective, thermal mass must be present in sufficient quantities and in the right places exposed to interior air.

Modern, climate-responsive green buildings, therefore, can use exposed thermal mass, in conjunction with cool outdoor conditions, to moderate internal temperature fluctuations and create a comfortable environment for occupants. Thermal mass has the ability to reduce energy use – by offsetting artificial cooling requirements – in climates where buildings need to be cooled during the day and heated at night. In addition, night ventilation can make use of a structure's exposed thermal mass along with low night-time temperatures to keep peak temperatures low in summer months.

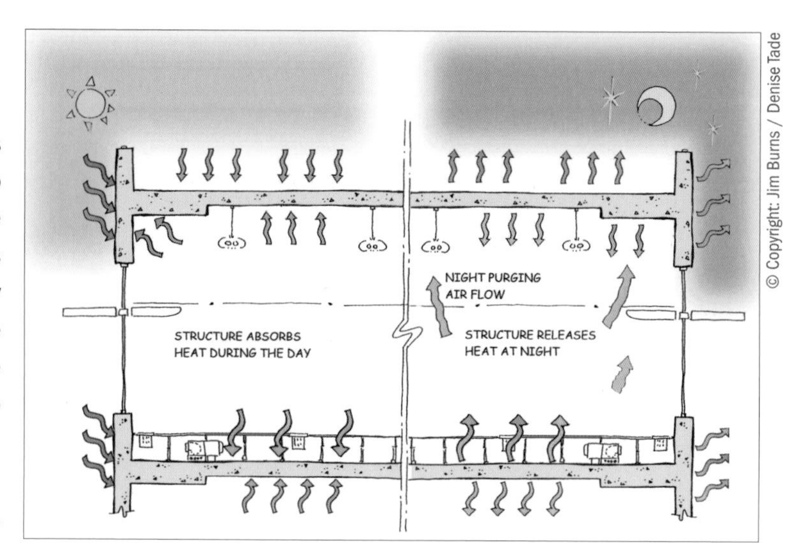

Thermal mass as a temperature regulator in buildings – for low-energy heat gain during the day and natural heat purging at night.

A Closer Look at Thermal Mass

Thermal mass is the ability of a material to absorb heat, using external energy to change the temperature of a high-density material such as concrete. This principle and technique has had an important role in low-tech buildings and yet, as with natural ventilation, its usefulness is misunderstood and often neglected when it comes to engineering design.

It is not uncommon for concrete and steel buildings to be designed the same way, despite the fact that they tend to create different interior conditions. Aside from a few passive solar homes built each year, engineers tend to view heat gain from the sun as an external environmental challenge. As a result, it is now much easier to find high-performing glass on the market that rejects heat rather than permits it. In general, thermal mass is conventionally viewed as "unreliable," and in a culture that demands instant gratification, the response times associated with thermal mass may appear too slow and potentially risky.

> Any third-rate engineer or researcher can increase complexity; but it takes a certain flair of real insight to make things simple again.
>
> – E.F. Schumacher, *Small Is Beautiful: Economics as if People Mattered*

More and more building science experts, however, are suggesting it is time to revisit the use of thermal mass. Studies are showing that the right combination of insulation and thermal mass materials, plus a building design that efficiently blocks even modest amounts of solar gain and heat, can help to curb peak cooling loads and further improve the energy performance of buildings.

For ecological engineers such as KEEN, using mass is fundamental to low-tech design. They see it for what it is: a tool with particular characteristics that, when properly used, can achieve superior low-energy solutions. Mass is not a panacea. It cannot do everything all the time, but it is highly useful and a key piece of the firm's sustainable design strategies.

KEEN's first real exploration into dynamic thermal storage came with the Library Square project in the mid 1990s, when designer Blair McCarry took advantage of the structure's exposed concrete ceilings and floors and underfloor air system to effectively condition the building. The firm has since incorporated thermal mass into a number of naturally-ventilated buildings as well – and its engineers continue to learn by monitoring the temperature performance of slabs and from further study of well-known European examples such as Lloyd's of London and the Ionica Building in the U.K.

Challenges/Opportunities

High-mass building materials and structural elements can offer significant energy benefits: shifting peak load conditions in order to moderate temperatures and therefore minimize or eliminate the need for mechanical cooling systems; allowing free pre-cooling in conjunction with night-time flushing; and absorbing solar gain to reduce winter heating or summer cooling requirements. These benefits are highly dependent upon a building's location and its design and operations.

One potential pitfall with this technique is temperature swings, by as much as 10°F, if the thermal mass as designed and constructed is insufficient to provide a cooling sink. Air temperatures also need to be managed carefully to maintain appropriate comfort ranges for occupants. Still, thermal mass design presents a tremendous opportunity for energy cost savings. A recent study reported in the *International Journal of Heating, Ventilating, Air-Conditioning and Refrigeration Research*[3] examined the potential for summertime utility savings at an Illinois field site using several thermal mass control strategies. Ultimately, the best-performing test strategy yielded a 40 percent reduction in total cooling costs as compared with night setup control. Representative utility rates for five locations (Boston, Chicago, Miami, Phoenix and Seattle) were used along with the models obtained for the field site. Significant savings were achieved in all locations except Seattle, where cooling loads were alredy low.

Best Practices

KEEN's experience has shown that thermal mass applications work best, in terms of realized savings, with lower capacity heating and cooling systems as part of a low-energy building design. The technique is also most appropriate for internal environments where clients and users can be flexible enough to allow temperatures to fluctuate slightly. When sized properly, thermal mass is able to moderate diurnal temperature swings effectively by absorbing excess heat during the day and unloading it at night in both heating and cooling seasons.

Concrete and masonry products are ideal thermal mass materials, especially those with a high capacity for heat storage, moderate conductance, moderate density and high emissivity. Typically, a three or four inch slab is sufficient for maintaining thermal comfort for people in buildings. Cost control is especially important with thermal mass, McCarry also noted, so that using the mass as a structural

image has been used twice - no others provided

Case Study

Evergreen State College – Seminar II Building

Evergreen State College near Olympia, Washington is known for its sensitivity to environmental issues, so it was no surprise that its leaders would want to pursue a green design for their new academic building. Seattle's Mahlum Architects was joined by KEEN Engineering and Tom Paladino as LEED® consultants on the project. Originally conceived as a single structure, the space was divided up into five distinct clusters made up of faculty offices, student "homerooms", seminar rooms, breakout spaces, a workshop, and a lecture hall. Known as Seminar II, the five 4-story buildings totaling 168,000 square feet are connected by a system of open walkways, stairs, and bridges that tie the levels together and provide increased seminar capabilities and support spaces for the college.

During planning, the project was marked by a great deal of client participation and consensus-based decision making. Thermal mass became an important component of the design, and its effects were carefully studied for the building. As a large facility with a wide variety of uses, it was not a foregone conclusion that passive, low-tech solutions could be utilized effectively, even in this Northwest climate. In the end, the project team did arrive at appropriate design choices where specific applications were found. As an example, many of the large classrooms could not harness natural ventilation exclusively because of high, widely-varying internal loads depending on whether a class was in session or not. Thermal mass was indeed too slow to react and the amount of exposed mass inadequate to keep large numbers of people comfortable. And yet, even in those spaces, operable windows were included so that mass could be tempered and play a role in energy reduction.

Still, with the site's average outdoor highs of 76°F during the summer, the use of natural ventilation as a cooling strategy was an obvious choice for the remaining portions of the building. Office spaces were designed to bring in outside air through operable windows and then flow through transoms above each door space. Smaller seminar rooms were treated similarly. The center spine for the building allowed air to flow up and out, and ventilation fans were added at the top of the building to ensure that airflow was good. As designed, the entire structural and exterior closure material is concrete – a system that on completion represented a building with at least a one hundred year lifespan.

Low-Tech Lesson V: *Do not assume that natural ventilation and thermal mass cannot play a role in large buildings. By breaking a large facility into more discrete elements, each can operate as a smaller, more passive facility.*

Low-Tech Lesson VI: *Passive solutions do not have to be "all or nothing." In buildings where some spaces are not appropriate for low-tech solutions, treat them differently rather than imposing their requirements on the rest of the facility. It is acceptable to have a building where parts of the facility are passively heated and cooled and others that are mechanically heated and cooled.*

or design element – and integrating it with other building envelope materials into the overall thermal load – is the best approach.

Thermal mass works well with natural ventilation – something KEEN discovered at York University's Computer Science Building and later the Seminar II Building at Evergreen College in Washington State. In buildings that rely primarily on natural ventilation for cooling, the mass can help to control diurnal effects by absorbing the heat of internal building loads on a summer's day and then purging that collected heat with cool air at night. Inside the building during the day, too, the thermal mass and operable windows work together as air-flows over the concrete structure to produce cooling air currents.

Design Considerations

As with natural ventilation and other low-tech strategies, the ecological engineer needs to be involved early – and preferably in the conceptual design stages of a project. "The earlier you get in the better," said McCarry, "because when you're considering thermal mass and natural ventilation or even radiant systems, it's critical to control the loads: the absolute value of the load as well as fluctuations of the load over a daily cycle." He added:

> It's also important to have simplistic solar control, building envelopes that neutralize perimeter effects and good control over interior loads to minimize temperature fluctuations over the course of a day. For example, if the sun is out, adding a little extra load, you might be able to minimize the use of electric lights and rely on daylighting. In short, you have to influence the control of loads – and you have to do it early.

Also important to include early in the process are simulations using thermal modeling, so that engineers are confident the building will operate as intended. Most energy modeling programs can not completely simulate the effects of thermal mass and so their benefits are easily ignored. Typically, systems are oversized to "overcome mass effects" rather than to harness them as the models do not

Case Study

York University Computer Science Building

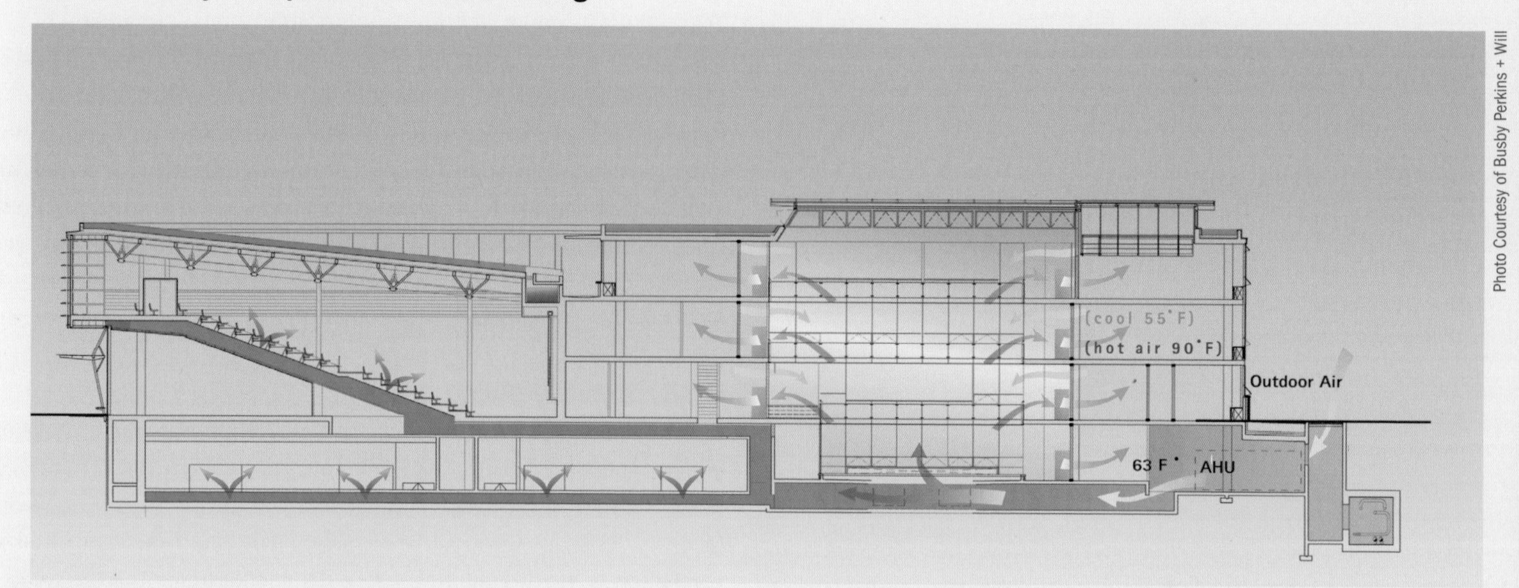

The Computer Science Building at York University by Busby & Associates has won numerous accolades for its innovative design and environmental performance as a cold-climate green building. Its success is due, in part, to the pivotal role of thermal mass in helping the facility utilize natural ventilation, even though Toronto's high summer humidity typifies a climate not usually considered for this approach. Like many naturally-ventilated buildings, the structure is composed of concrete; during the cold weather months, much of this mass is used to temper interior temperatures as winter sun streams into the building.

Also of interest was the project's creative use of the ground as thermal mass. Air is brought into the Computer Science Building through extremely long "earth tubes" (almost four hundred feet in length) buried beneath the building. As the air is introduced into the tubes, the temperature of the surrounding earth passively conditions it. In summer, the earth is much cooler than the outside air temperatures, so incoming air can be tempered and even partially dehumidified. The tunnels are sloped so that condensation flows out and away, and potential air quality problems do not occur inside the tubes and building. The earth tubes effectively lengthen the season in which low-tech solutions can be utilized. Ultimately, the natural ventilation and thermal mass scheme in use at York's award-winning building has saved the University more than was originally predicted from the computer simulations. In fact, the estimated net annual consumption of energy for heating, cooling, and electrical demands exceeded the original target by 15 percent. The simulations predicted an annual energy use of 1940 MJ/square meter (30 percent better than ASHRAE 90.1 requirements), and actual consumption during the first year of operation was 1,644 MJ/square meter.

Low-Tech Lesson VII: *Even climates that are not considered ideal for low-tech solutions often have shoulder seasons where the strategies will work. Consider designing hybrid systems that can work in passive mode for parts of the year.*

Low-Tech Lesson VIII: *Use earth coupling wherever possible as part of a low-tech solution. This passive strategy can manifest itself in solutions such as ground-source heat pumps or as earth tubes to pre-temper air. Remember that the earth is usually cool when needed most and warm when needed most.*

take into account the benefits of thermal lag. However, the ecological engineer designs mechanical systems to harness the beneficial effects of thermal mass.

Another point to consider is the unpredictability of thermal mass. It is not uncommon, considering climate and time of year, for the heating or cooling cycle to carry over into the next day. If heat is not completely purged out on a given night or if an especially cool evening causes the mass to be overcooled, building operators could face a problem the next day. Most standard North American load programs do not accurately address thermal mass performance, and many of the calculations will apply a two- or three-hour time delay to portions of loads such as lighting. Also, a building that combines thermal mass and natural ventilation should include a control sequence to monitor temperatures – and monitoring should be conducted over a broader timeframe than usual: think in terms of days, not hours, to account for carryover and the natural cycles of thermal mass. As a result, night setback is not recommended, because the thermal inertia of changing the temperature of the mass by several degrees could translate into a day or two – or more – of bringing the building back into line.

Further Thermal Mass Resources

Passive and Low Energy Cooling of Buildings (1994), Baruch Givoni, published by Van Nostrand Reinhold, New York, covers the role of thermal mass in night-time ventilating and radiant cooling systems and the earth as a cooling source.

Sun, Wind, and Light: Architectural Design Strategies (1985), G.Z. Brown and Virginia Cartwright, published by John Wiley & Sons, New York, NY.

Part Three

Like a Second, Or Third, Skin

> Life set itself to new processions of seed-time and harvest, the skin newly turned to seasonal variations, the very blood humming to new altitudes.
>
> – Mary Austin

A building's façade is truly its skin: the more efficiently it breathes, aborbs, protects and regulates temperature as needed, the better it functions – sustainably – over time. Sunshades, overhangs, louvers, awnings, trellises as well as high-performance cladding, glazing and roofing all play an important part in maintaining comfort indoors. The physical interface (envelope) is equally important in determining the quality of those interior spaces and how well the building interacts with the outside environment.

Because the façade is acted on relentlessly by conductive, convective and radiant forces, low-tech buildings live or die based on how well their exterior elements are designed and maintained. These design features are typically the domain of the architect, but the ecological engineer must also understand and contribute to the design in order to produce a highly efficient building. KEEN's experience, again, is that low-tech solutions only work when a building is considered as a complete system – from outside to inside and inside to outside.

A Closer Look at Façade Integration

The use of double-skin façades is more pronounced in Europe, driven in large part by architectural preference for all-glass exteriors – but also for the practical tie-in with natural ventilation schemes, since single-layered façades have more acoustic and security constraints. It has yet to catch on completely in North America due to the real and perceived additional first costs of applying this façade design, although it is offset to some degree by energy and operating cost savings.

Still, KEEN's track record with this low-tech approach keeps growing for both new and renovated structures. Vancouver's revitalized Telus/William Farrell Building became the first double-glazed, triple-skinned building in North America when it reopened in 2000. For the new Revenue Canada building in Surrey, British Columba, KEEN engineers collaborated with the architect Peter Busby to include sunscreens, light shelves and clear low-e glazing to enhance the penetration of natural light, filter solar gain and reduce operating costs. And for the BC Gas/Terasen Operations Centre, KEEN designers recommended significant exterior shading on the south exposure and massive architectural walls with punched windows for the east and west sides of the building, using "sun shields" for solar control.

With these and other recent projects requiring smart façade integration, KEEN designers have learned the importance of early collaboration with architects. This dialogue is essential to informing how the building will react to climate and to understanding the true role of the façade in overall performance. The right input and integration at the right time can save considerable amounts of energy.

Challenges/Opportunities

Greater use of double-skin façades in Europe, especially for the upper stories of naturally-ventilated high-rise buildings, has been given some study by the engineering community. In *Architectural Record*, authors Lang and Herzog reported that occupants of these buildings with "twin-face" glass façades are more satisfied and productive than in conventional buildings – as a result of the natural ventilation, daylighting and greater environmental control over their workspaces[4]. And while these façades are more expensive than conventional curtain walls in Europe – and an even higher expense in North America – the article cited estimates by engineers and owners that certain ventilated façade types were instrumental to energy savings of 30 percent to 50 percent.

A growing volume of literature on dual-layered/double-skinned façades is examining this technique's overall performance in terms

Case Study

Revenue Canada

© Copyright: Jim Burns

When Revenue Canada decided to build a new facility in Surrey, British Columbia, it chose a design-build process supported by a collaborative team to include the builder as well as the architects (Busby Perkins + Will) and engineers. The agency's desire was a high-performance building that represented good fiscal judgment: one that saved taxpayers money over time and would not be viewed as a waste of funds upfront. In this project, the design of the façade was very important for regulating loads – but energy-saving ideas were carried inside the building as well, where decisions were made in the planning process to further improve the building's performance. All offices were moved from the perimeter of the building (per standard practice) to the interior. This move reduced the need to treat the perimeter of the building differently, while incorporating an underfloor air distribution system and allowing for greater daylight penetration. Lighting controls that respond to daylight levels saved a great deal of energy. In addition, the building façade incorporated shading devices to reduce glare and heat gain, and exposed concrete on the interior helps to regulate temperature fluctuations. Each component became an important part of the overall building system, and each team member contributed in different ways to meet the project's goals.

Low-Tech Lesson IX: *Low-tech solutions do not have to cost more upfront than traditional systems and, in fact, sometimes cost less. Most of the time, there is a cost transfer and, as some components such as mechanical systems get less expensive, other areas like the façade get more expensive. The trade-offs should be viewed together, not in isolation.*

Low-Tech Lesson X: *Sometimes interior solutions can affect the performance of low-tech strategies. The choice of interior paint color, for example, can greatly impact the effectiveness of daylight and improperly designed lighting controls can nullify savings that might exist. As stated in the last two chapters, the building must always be viewed as a complete system.*

Case Study

Telus/William Farrell Building

The old 1960s-era Telus/William Farrell Building in Vancouver was recently rejuvenated to create a new "look" for the building and improve its performance and comfort. The original façade of the building – featuring single-pane windows and low levels of insulation – had been deteriorating. The interior was often uncomfortable and the outside drab and uninteresting. All of this soon changed as KEEN proposed a double façade system that would operate as a second skin to the facility, rather than simply updating the mechanical system.

© Copyright: Jim Burns

Double-glazed glass panels were suspended from the exterior to surround and insulate the building. Operable windows were located on both skins. In winter, the building is buttoned up. Both sets of windows are closed and the air space between the façades acts as a thermal buffer, greatly reducing winter heat loss and creating a warm layer of air when the sun is out. During swing seasons, the façades operate in a variable mode in response to outside air temperatures. People are trusted to open and close the interior windows as desired and, as with other KEEN buildings, they are given basic instructions on when the windows can or cannot be opened. In summer, the outer layer of windows is kept closed; even when people open their interior windows, air transfer is limited.

Low-Tech Lesson XI: *Low-tech solutions always begin with the building envelope. Make sure envelope decisions are closely coordinated with the design of interior systems. Reducing loads at the perimeter is always the most effective first step.*

Low-Tech Lesson XII: *Find multiple reasons for each design solution. In the Telus building, the new façade created a more comfortable, energy-efficient solution, but it also satisfied the client's wish to update the exterior image of the facility. Ideas that work in multiple ways are seldom value-engineered out.*

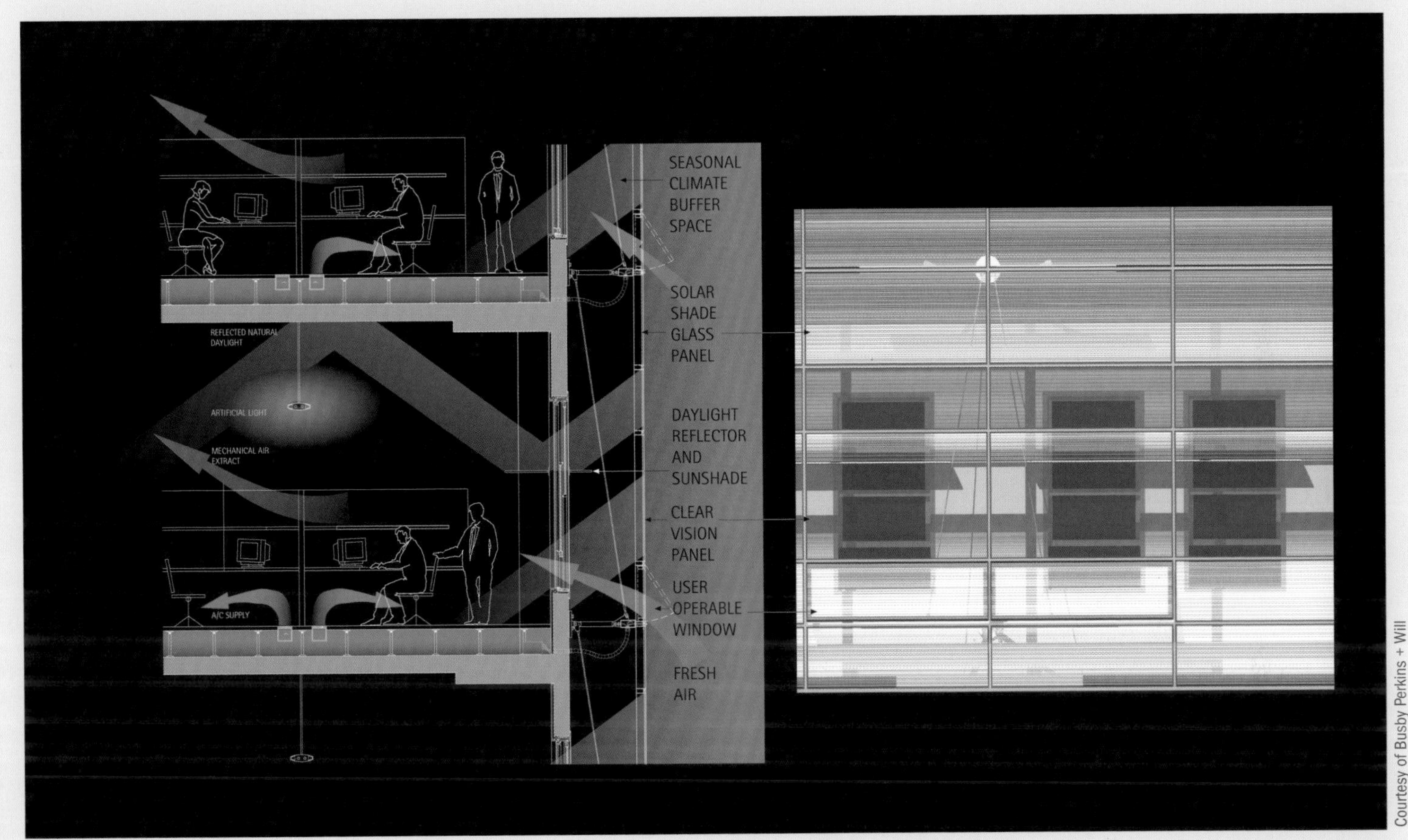

Courtesy of Busby Perkins + Will

Diagram demonstrating dynamic interactions of sunlight, heat gain, and comfort control in the Telus/William Farrell Building.

of acoustics, airflow, thermal comfort, energy use, and daylight as well as moisture and fire protection. There are concerns about a lack of reliable simulations and practical measurements but, in general, double façades offer the advantages of acoustical insulation against outside noise, provide efficient cooling by ventilating summer heat through a façade's air gap, and give occupants the ability to open windows even in high-rises. Double façades do pose a few challenges as well. The added glazing can restrict the amount of daylight entering the building, so material choice and design are critical. Also, because the inner surfaces of outer panes produce condensation, they may require frequent cleaning.

Best Practices

The façade's fundamental purpose is to optimize the thermal performance of a building. Careful consideration should be given to the design and placement of interior and exterior shading devices, the size of openings, the thermal integrity of the building envelope, and use of insulation to reduce heat losses or gains.

A double-façade design is essentially two glass skins separated by an air space that ranges in width from 6 to 30 inches. To make this space an active air plenum, motorized dampers may be placed at the bottom and top of the space, with fans at the top to assist in air movement. These dampers are open or closed according to season. The primary skin consists of a conventional or massive wall system with an outer layer of single-glazed panels. The secondary skin, then, is a non-load-bearing curtain/solar glazing system that utilizes a combination of opaque and clear vision fritted glass.

Selection of glazing is more critical and varied than ever, as new types of glass are dramatically improving the possibilities of building energy performance by influencing heat loss, solar heat gain, daylight and visual access to the exterior. The practice focuses on several key criteria:

- solar heat gain coefficient (SHGC) and shading coefficient (SC), according to a project's solar control strategy for maximizing heating or minimizing heat gain
- overall U-value, particularly for cold climates, in minimizing energy consumption and reducing the need for perimeter heating
- daylight potential, or visible transmittance (VT)
- translucency/opaqueness, depending on the amount of available shading, so occupants can see outside the building

Finally, designers should pay close attention to the thermal resistance of walls and roof assemblies in meeting performance standards. Avoid thermal bridging by using non-conductive fasteners or adding thermal breaks. And the building's envelope air/vapor barrier should be continuous, able to support wind load or pressurization without displacement, and located for servicing as necessary.

Design Considerations

In designing the façade system – whether single-, double- or triple-skinned – it is important to first address occupant comfort including mean radiant temperature, lighting and glare issues. "That's the first step in making it a better building," said McCarry, "and that goes beyond dry bulb temperature. We can do a lot with façade to increase comfort."

Engineers should also suggest performance targets for the building based on their system design. Particularly with a double-skin façade system, airflow modeling – using the mass balance network method or computational fluid dynamics – is useful in design decisions. On the Telus project, KEEN conducted a number of modeling studies using TAS (Thermal Analysis Software) to determine the effectiveness of the air plenum.

Often overlooked in North American façade design are perimeter zone comfort and radiant energy issues. "Part of that is we don't have a lot of good tools for measuring comfort," said McCarry. "We play with mean radiant temperature because of thermal mass and chilled slabs. But most engineers are not actually aware of mean radiant temperature issues, and it's not usually taken into account in our designs."

Given the increasing number of sophisticated glazing choices available, there are several ways to improve window performance. Specify window frames that enhance overall thermal resistance and ensure minimal infiltration and leakage. Minimize solar gain through sun shades and window treatments such as fritted glass (opaque screen) panels. And factor in the various orientations of a building and the SHGC and SC characteristics of the glazing according to east, west, north and south exposures.

Conclusion

If there is one dominant trait shared by KEEN designers and the firm overall it is: attitude. As "natural thinkers", their passion and their pursuit of low-tech building solutions is defined best as a drive for simplicity, optimization and elegance in mechanical and electrical design. It is fundamental to engineering and so represents a return, of sorts, to the idea of First Principles – to creating systems that do the most amount of work with the least amount of energy and materials. Likewise, historical precedence is important, "going back to the old ways of doing things" when and where appropriate with natural ventilation, daylighting and thermal mass.

Low-tech design practices at KEEN, too, are strengthened by focus – by a consistent, unwavering focus on a specific mechanical problem or design challenge without marginalizing the solution. Instead of making minor improvements over time, this focus requires an approach that understands and addresses the essential needs of a building, new or renovated. And it results in smaller, more efficient structures that use half the energy at half the expense. Most importantly, these structures become people-friendly spaces. Buildings that celebrate air, light and comfort. Places with soul.

Further Façade Integration Resources

"Energy Performance Assessment of Multiple-Skin Façades", D. Saelens, Ph.D., J. Carmeliet, Ph.D., H. Hens, Ph.D., *International Journal of HVAC&R*, April 2003. This paper discusses modeling the energy performance of an office equipped with a conventional insulated glazing unit and exterior shading and with three multiple-skin façade typologies under typical Belgian weather conditions.

Glass Construction Manual (1999), C. Schittich, G. Staib, D. Balkow, M. Schuler, W. Sobek, published by Birkhauser, Munich, explains some of the fundamentals of double-skin façades as well as other advanced façade systems.

"Performance of Solar Façade Components", International Energy Agency (IEA), www.iea-shc-task27.org. A function of the IEA Solar Heating & Cooling Programme, this research task is designed to assess the solar, visual and thermal performance of dynamic glazings, daylighting products, solar protection devices, and double-envelope systems.

"Study of Current Structures in Double-skin Façades", S. Uuttu, published by the Helsinki University of Technology, www.glassfiles.com/library.

High-Tech Solutions

Water, Air and Radiant Heating/Cooling

Technology, like art, is a soaring exercise of the human imagination.

– Daniel Bell, "Technology, Nature, and Society," *The Winding Passage*

Equally as important to mechanical innovation and excellence in sustainable design are KEEN's "high-tech" solutions for comfort and water and energy efficiency. In contrast to techniques that rely on natural ventilation or thermal mass, KEEN's high-tech approaches are dynamic, active and technology-based, each relying as much on precedent, iterative processes and modeling as on successfully educating clients and code authorities. Again, KEEN's team members have learned much from leading European examples for ventilation or radiant heating and cooling, adapting them to North American standards and, in many cases, discovering ways to improve on the designs. But they have also tapped some of the most advanced water and wastewater technologies in the world, this time from the U.S. and Canada.

As ecological engineers, KEEN effectively utilizes water, air, light, earth, concrete and other elements to enhance the performance and cost-effectiveness of its MEP systems – always keeping the simplicity of designs and control schemes in mind, integrating with other systems and disciplines where possible. This chapter briefly explores multiple design techniques employed by KEEN engineers to create high-tech solutions such as:

- Water Conservation Methods
- Rainwater Collection/Re-Use
- Biological Waste and Wastewater Treatment
- Underfloor Air Distribution
- Displacement Ventilation
- Radiant Heating and Cooling Systems

Part One

The Strategy of Water

Building construction and operations rely heavily on potable water from local and regional environments. U.S. buildings alone use 17 percent of available fresh water annually, according to the Worldwatch Institute. In many parts of North America and the planet, fresh water is an increasingly scarce resource. Even water-rich regions like the Pacific Northwest have been affected where, as recently as 2003-2004, the cities of Vancouver and Victoria faced severe water restrictions due to below-normal rainfall during the previous winter. Growth in urban water use is lowering water tables, while the costs of obtaining and treating new sources of water have steadily risen, making demand-side options economically attractive.

> Since the turn of the century, worldwide municipal use of water has grown 19 times and industrial use has grown 26 times. In contrast, agricultural use has increased only 5 times.
>
> – Worldwatch Institute, 1999

In addition to the expense of obtaining and purifying water, sewage distribution costs are mounting. Many older North American cities still rely on sewage facilities providing only primary levels of treatment; others, like Vancouver, still predominately utilize "combined" systems – those with a single pipe for both storm and sewer flows. In the face of rapidly-growing costs to upgrade out-dated infrastructure and sewage plants, municipalities now focus increasingly on sustainable ways to lower water consumption, control stormwater, and treat or recycle wastewater on site.

A Closer Look at Water Efficiency

By some estimates water conservation and re-use measures in commercial buildings can easily reduce water usage by 30 percent or more. In a typical 100,000-square-foot office building, low-flow fixtures coupled with sensors and automatic controls can save a minimum of one million gallons of water per year, based on 650 building occupants each using an average of twenty gallons per day.

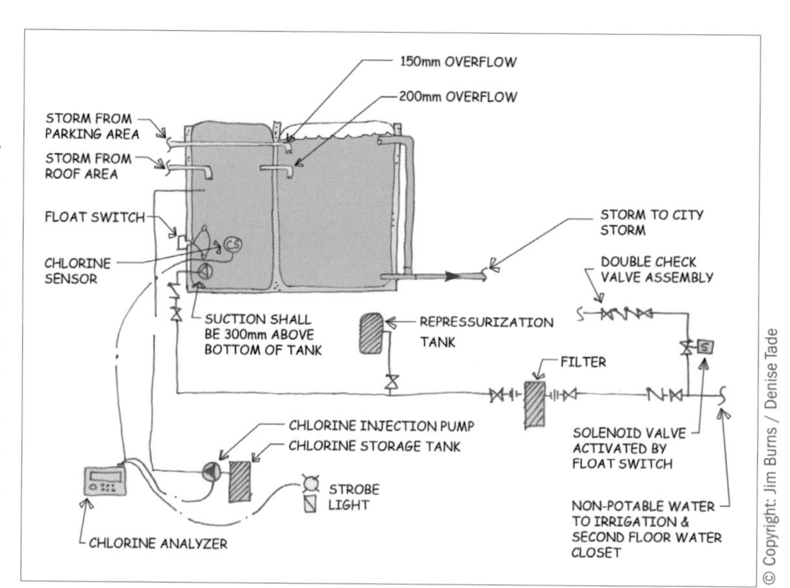

Sample KEEN stormwater storage and treatment system: Modified to suit each project's site conditions, water quality and building demands.

In many urban areas, residents are discovering that steady improvements in water-conserving technologies, along with innovative management practices for toilets, showerheads, irrigation equipment and faucets result in unchanged or even improved levels of customer service.

For plumbing engineers at KEEN, water conservation and re-use methods have represented a return to design basics, a philosophy of keeping processes simple. Mark Swain, an associate in KEEN's plumbing department for twenty-six years, pointed out:

> Throughout the 1980s and 1990s, we got really technical with our systems. Now we're getting back to simpler designs. For us in plumbing, it's a matter of re-using things: blackwater systems,

greywater systems, storm retention and other applications that make common and economical sense for buildings.

Challenges/Opportunities

Using large volumes of water clearly increases maintenance and lifecycle costs in building operations. Conversely, reductions in water use and improved water efficiency can save thousands of dollars per year in energy and other operating costs. Water conservation can also lead to lower water rates, more stable municipal taxes and a reduced burden on local water processing facilities.

The collection and reuse of rainwater has been a standard practice in many parts of the world for decades. Rainwater harvesting offers an opportunity for buildings to offset water needs and reduce the on-going expense of main-supplied water – as well as providing a backup supply during periods of water restrictions.

KEEN has found there are consistently two real barriers to acceptance of rainwater harvesting and certain water conservation measures. Number one is codes – gaining health authorities' approval on alternative approaches such as waterless urinals or rainwater use in toilet flushing. The second barrier is economic. In most locales, water is cheap, so clients may be skeptical about additional capital expenditures. The use of low-flow fixtures is cost neutral; however, a more progressive measure like rainwater collection and storage can represent an initial project cost premium.

Best Practices

First, KEEN engineers recommend water-conserving measures on every new project, specifying fixtures such as ultra-low flow faucets with aerators and automatic shut-off sensors, low-flush or dual-flush toilets, low-flush or waterless urinals, and water-saving showerheads. Other water conservation methods considered include: push taps, self-closing flow controls and tap flow regulators for hand wash basins; cistern catchment devices; and urinal controls (infrared, radar, autoflush). KEEN next investigates opportunities for rainwater collection/re-use and greywater systems to supply irrigation and/or toilet flushing. In rainwater harvesting, preferred roofing materials are metal, clay, or concrete-based, using gravity flows to divert water into rain barrels or cisterns. Storing roof rainwater in cisterns does require filtration and some form of treatment for algae. Through effective integration of stormwater strategies at the White Rock Operations Centre, for example, building design made possible a reduction in potable water use of 88 percent annually.

A fundamental first step on projects is investigating the possibility of water conservation through stormwater management and re-use schemes (White Rock Operations Centre).

KEEN also attempts to closely integrate all aspects of a project's water strategies with on-site stormwater management measures.

These measures include:

- Impervious surfaces with pervious paving to absorb excess stormwater runoff (especially from parking areas)
- Small surface and subsurface infiltration basins, bioretention cells and vegetated swales
- Xeriscaping with native plants to significantly reduce water needs
- Use of "green roofs" to further reduce heat gain and heat loss

Design Considerations

In response to the overly complicated designs of the past, Swain noted that KEEN now works hard to rationalize every detail of new plumbing systems. In many integrated care facilities, for example, designers typically installed flush valves in water closets; now they specify dual-flush systems, which translates into less water and smaller piping systems.

Particularly with stormwater, it is important to check with local health departments for guidelines regarding rainwater collection since it is not federally regulated. If harvested rainwater is intended for potable or irrigation purposes, certain health codes may require back-flow prevention devices to avoid the risk of contaminating public drinking water supplies.

A Closer Look at Bio Waste Treatment

Whenever possible on projects, KEEN Engineering explores the feasibility of using biological methods to treat liquid waste on site and within its buildings to reduce the volume of blackwater entering municipal sewage systems. Often, these treatment alternatives are capable of producing high-quality water for multiple uses, yet they also call for the engineers to navigate carefully through the regulatory approvals process in most jurisdictions in Canada and the United States.

Case Study

Calgary Water Centre

Water conservation and stormwater management were essential considerations in the design of the new Calgary Water Centre with Manasc Isaac Architects, a new 183,000-square-foot office building for the City of Calgary, Alberta. In addition to sustainable features like in-slab floor cooling, underfloor air supply and 100 percent daylighting, KEEN engineers specified an array of low-flow water fixtures plus waterless urinals and a rainwater collection/recycling tank for use in irrigation and flushing toilets.

Slated for completion in 2006, the Centre will house approximately four hundred Water Works and Wastewater department staff on three floors of open office space overlooking a central atrium. With a goal of achieving LEED® Silver, KEEN estimates a 59 percent reduction in water use. Engineers have also recommended on-site wastewater treatment, using constructed wetlands sized for the building's entire sewage load and water reuse as a demonstration project.

Gaining approval on stormwater use in toilets was an important move on the project. Calgary's plumbing code is specific about allowing only the use of potable water for this application. On a recent residential development in Calgary, KEEN had met several times with the regional health authority to address the authority's concerns. These meetings resulted in several small changes to system design. With a letter of approval in hand, KEEN's Jim Sawers was able to convince the city's plumbing inspectors of stormwater use viability for the Water Centre as well.

The initiative, on KEEN's part, was simply investing the time to meet with various parties, to educate them on this approach as a sensible and sustainable one – and then get their consent. This protocol sets a precedent for what needs to be done to change the code.

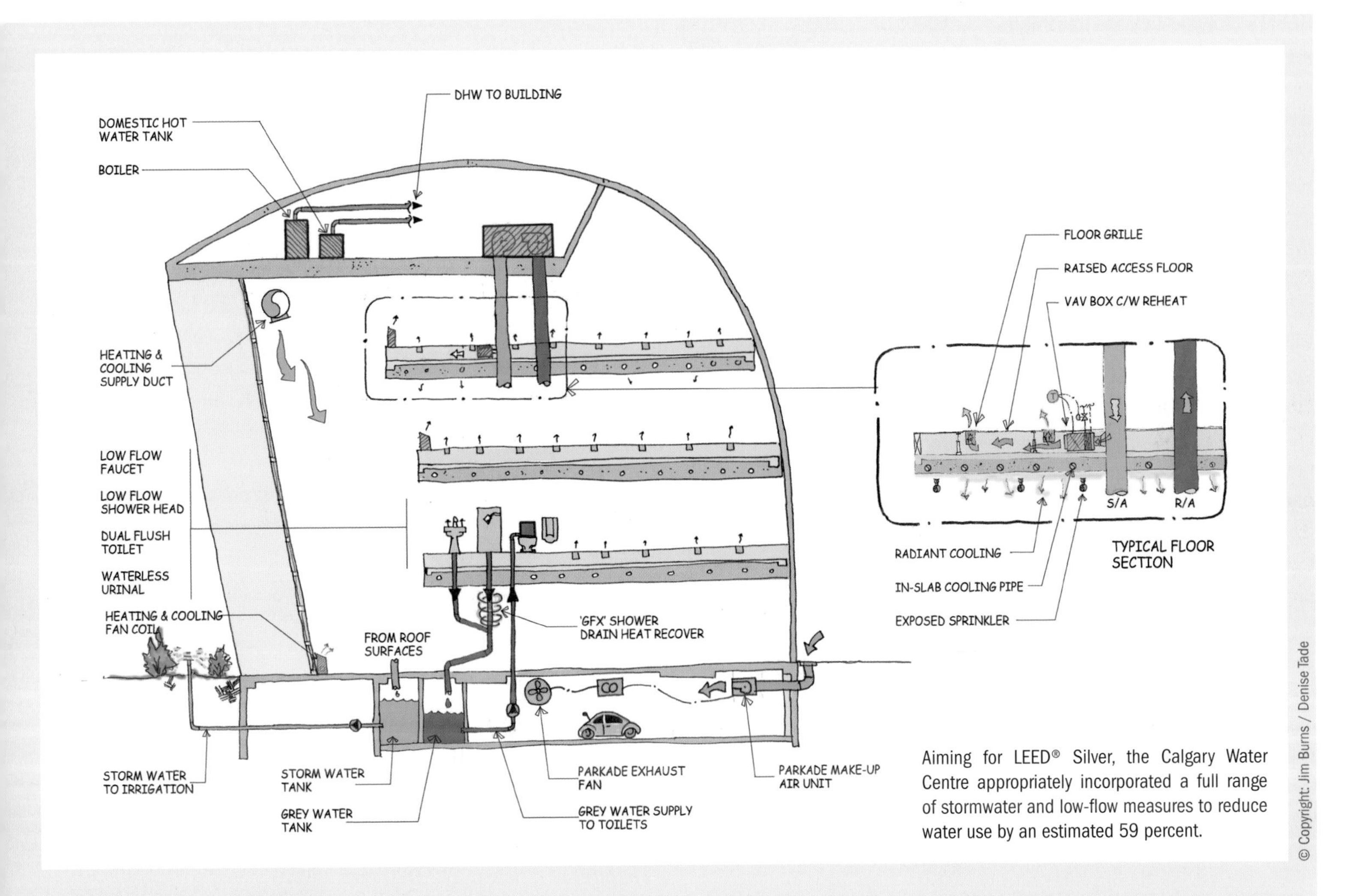

Aiming for LEED® Silver, the Calgary Water Centre appropriately incorporated a full range of stormwater and low-flow measures to reduce water use by an estimated 59 percent.

High-Tech Lesson I: *Potable water is used most efficiently for applications where a person ingests water, such as drinking fountains, sink faucets and showers. Non-potable water, such as greywater or collected rainwater, may then be cost-effectively diverted for use in landscape irrigation, toilet flushing, custodial purposes, and other building systems.*

High-Tech Lesson II: *Beyond the fundamentals of low-flow fixtures and rainwater collection, sustainable engineers are in a strong position to help a project team explore additional water conservation measures for a building site like pervious paving surfaces, bio-swales and xeriscaping.*

KEEN's biological waste treatment installations have included constructed wetlands, aerobic treatment systems, hydroponics-based systems (Living Machines™), and composting or ecologically-based toilets. For the new $500 million Vancouver Convention Centre expansion, to be completed by 2008, KEEN will install a blackwater treatment system (approximately 10,038 gpd) and a wide variety of water conserving, recycling and re-use solutions. Recently, engineers also contributed their expertise to a multi-stage wastewater treatment system (66,043 gpd) at the Columbia Icefield Visitors Centre, located in Jasper National Park, Alberta.

> Education on a lot of these new water and wastewater technologies is really essential. It's just a matter of time and process and teaching our children. And a lot of the dialogue we have with architects, building owners and others is to simply make them aware that it's not really a drastic change in one's life. Embrace it.
>
> – Mark Swain, KEEN

Living Machines™ – natural wastewater treatment and reclamation systems – have been proposed for the new Convention Centre as well. In addition, KEEN engineers were instrumental in designing the plumbing systems for Living Machine™ installations at The Body Shop-Canada headquarters (processing 4,000 gpd) in Toronto in 1993 and the IslandWood educational facility on Washington State's Bainbridge Island (2,400 gpd) in 2002.

KEEN's experience with composting toilets dates back to the C.K. Choi Building, where UBC's already overloaded sewer system would have required the University to upgrade the local sewer main at a cost of more than $100,000. Not only was the $70,000 premium for composting toilets less expensive, water resources were further conserved. The composting toilets, located on all three of the research center's occupied levels, save nearly 100,000 gallons of water per year. Greywater is used to irrigate the surrounding landscaping.

Case Study

Dockside Green

KEEN Engineering is playing a key role in Dockside Green, a residential/business development now underway on an 11.6-acre brownfield site located in the heart of the City of Victoria, British Columbia developed by Windmill/Van City Enterprises and designed by Busby Perkins + Will. The master plan for this new mixed-used, sustainable community has targeted LEED® Platinum, with progressive features to include biomass energy cogeneration and complete on-site greywater and blackwater treatment. In addition to water reduction in all buildings, the development will use only non-potable water for irrigation and other community uses.

Treating all water and sewage on site is critical to the entire development. Dockside Green's self-contained treatment system will utilize bioreactor tanks and ultra-filter membranes to treat all of the sewage as high quality water for unlimited re-use classification (toilet flushing, irrigation). This treatment will be followed up by activated charcoal and UV filters, removing all bacteria, viruses, metals, nitrogen, phosphorus, and other contaminants to produce water quality that exceeds any of the outfall requirements of Canada's Ministry of the Environment. Two bio-swale systems will be in place as well: one for discharging treated water into Dockside's stream; and the second for stream overflow discharged into the harbor. Initially, this tertiary treatment system will process 150,000 gpd.

Dockside's treated water will then be available to flush toilets and to provide irrigation on green roofs. This on-site treatment of sewage will also assist developers in creating a closed-loop cycle where the treated water reduces operating costs and provides water to the landscaping and pond features, further increasing the livability and marketability of the development.

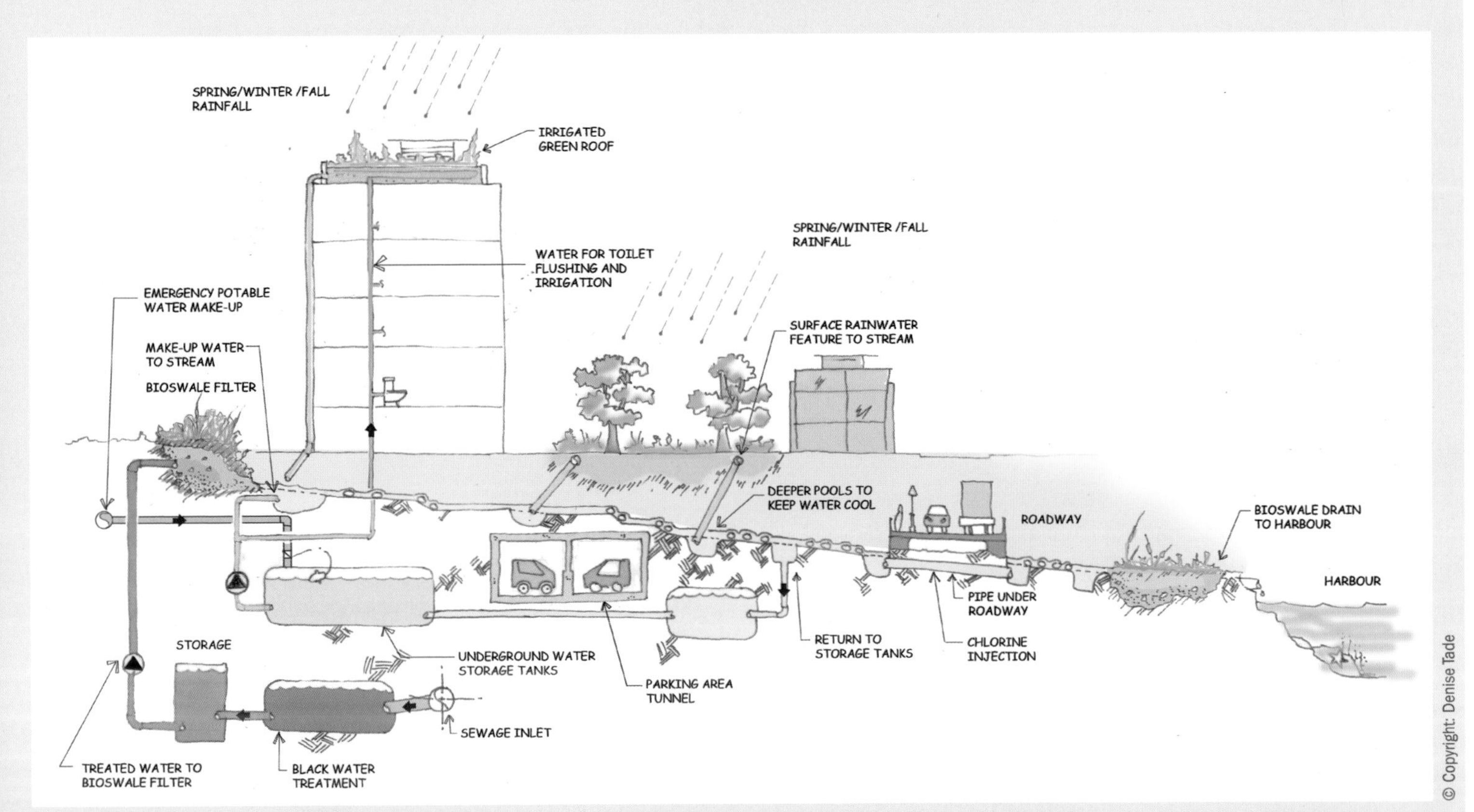

State-of-the-art blackwater treatment – using bioreactor tanks and ultra-filter membrane technology – will be a central feature of the new Dockside Green residential/commercial development (Victoria, BC). Treating sewage and greywater on site provides water for irrigation and landscaping features as part of a closed-loop cycle.

High-Tech Lesson III: *Develop appropriate design strategies and select an appropriate wastewater system based on each building's program, occupants and site – and for its ability to integrate with other related treatment applications.*

High-Tech Lesson IV: *Always check with the local health authorities for regulations governing the use of greywater and blackwater systems and the permits required. Every state and province has its own standards for greywater irrigation systems.*

Challenges/Opportunities

Each of these on-site wastewater treatment systems is designed to conserve water and transform "wastes" into usable resources. Diverting the greywater, blackwater and solids away from a local wastewater treatment plant also reduces costs related to wastewater surcharges, energy use, sludge disposal, and chemical treatment. There are other benefits, too, in the case of Living Machines™ – the systems can offer significant opportunities to educate local communities and dramatically increase awareness of sustainable practices. Gaining initial buy-in from clients and code authorities for the Choi's composting toilets, IslandWood's Living Machine™ and, more recently, an advanced membrane waste treatment technology for the Dockside Green development and Vancouver Convention Centre, has been perhaps each project's biggest accomplishment.

Swain and others have found that the first major obstacle for clients in accepting green plumbing applications is often money – again, a mind-set that these applications are more expensive than conventional approaches. Convincing local municipalities and health authorities about the merits of alternative systems has been a second major obstacle.

"With a lot of the water and waste technologies, it's emotional, the stigma of doing something different," said Swain. "People can readily understand how a composting toilet works, but they're not sure they want it – the same with waterless urinals. So our challenge then becomes how to help people get past those barriers."

Best Practices

New on-site wastewater treatment technologies continue to emerge, such as aerobic biological treatment reactors, and they will undoubtedly have a central role at Dockside Green and other upcoming developments. Meanwhile, constructed wetland systems remain a significant and sound solution, both environmentally and economically, in treating both greywater and blackwater as part of a permanent installation. A constructed wetland mimics the way waste is processed by a natural riparian community, using microbes and plants to break down waste. Common wetland system options include: surface-flow wetlands, which consist of a tiered system of ponds filled with wetland plants; and subsurface-flow wetlands, which use a gravel medium to anchor plants instead of soil.

Living Machines™ have been put in place in communities, institutions, resorts and other facilities across North America over the last decade. These ecologically-engineered systems use a "natural-systems" approach to treat wastewater, using microorganisms, small invertebrates (such as snails), plants and sunlight. Capable of producing 600 to 750,000 gallons per day, depending on the application, the final, high-quality effluent is suitable for a variety of reuse applications and will generally meet tertiary standards (BOD, TSS, and TN) of about 10 mg/l. Specific components are selected by designers according to the characteristics of the wastewater to be treated and the treatment objectives.

Composting toilets, although not as widely used in commercial or institutional buildings, can be applied in nearly all climates. Available as either self-contained units or central systems, these installations are designed to separate liquids from solids in toilet waste. Then, a natural process of heat and fresh air stimulates bacteria in the solid waste, breaking it down into an odorless, nutrient-rich humus. The finished compost is periodically removed from the unit. Local regulations will vary on its proper disposal; in some jurisdictions it can be used as landscape fertilizer.

Design Considerations

So why recommend alternative bio-waste treatment if there is relatively easy access to a municipal sewer system? For Dockside Green, Victoria's sanitary system is dumped straight into the ocean, which is not responsible and so the new development will serve as a regional model of sustainability. Over top of the new Vancouver Convention Centre will be a 280,000-square-foot living roof – the largest in Canada – requiring regular irrigation, particularly during the city's dry summers. At the Choi building, composting toilets provided

© Copyright: Jim Burns

The C.K. Choi Building's subsurface "biological marsh" was conceived to treat greywater and composting liquids for effective use in irrigating the surrounding landscape.

an ideal alternative where a sewer extension on campus would have been cost prohibitive – and KEEN still was required to include a sanitary connection as a future contingency.

For constructed wetlands, it is important first to identify design requirements based on users, capacity, pollutants to be removed from water, as well as the amount of time necessary for thorough treatment, and vegetation and aquatic life survival requirements. Surface and sub-surface flow wetlands are possible in almost any climate; however, they may require significant landform modification adjacent to the unit. On large Living Machine™ projects, the primary cost consideration is providing adequate space for the process itself and a greenhouse structure for reliable operation during periods of cold weather in more temperate climates.

With composting toilets, climate can be a factor; for example, particularly when temperatures drop below about 55°F, the bacteria that process the waste in a composter slow down their activity. Therefore, in colder climates, additional equipment such as solar heating may be necessary. Maintenance and operations of these systems is relatively simple. One end-product, the "composting tea," may be diluted and combined with collected greywater to irrigate landscaping through a subsurface system. Photovoltaic-powered fans may also be used to provide additional ventilation in units.

Further Water and Waste Design Resources

"Water Reuse Standards and Verification Protocol" (December 2004), a research report from the Canada Mortgage and Housing Corporation (CMHC), examines current regulatory standards in use around the globe.

"Rainwater Harvesting and Grey Water Reuse" (March 2003), from CMHC, reviews practices involving water reuse in residential and other buildings and looks at regulations and standards governing non-potable water.

"Cisterns/Rainwater Harvesting Systems", produced by British Columbia's Capital Regional District Water Department, can be found at www.advancedbuildings.org/main_t_plumbing_cisterns.htm.

"Water Strategies," Minnesota Sustainable Design Guide, is available at www.sustainabledesignguide.umn.edu/MSDG/water.html.

"Guiding Principles for Constructed Treatment Wetlands: Providing for Water Quality and Wildlife Habitat" (October 2000), published by the U.S. Environmental Protection Agency, includes information on siting, design, construction, operation, maintenance, and monitoring of constructed treatment wetlands.

"Regulatory Barriers to On-Site Water Reuse" (1998), produced by CMHC, examines national, provincial and municipal regulatory barriers to implementation of on-site water reuse technologies – and states how they can be overcome through provincial codes that allow innovation.

Part Two

Air Alternatives: Underfloor Air and Displacement Ventilation

Underfloor air distribution (UFAD) and displacement ventilation (DV) techniques continue to attract greater interest among engineers and architects. Each represents an innovative alternative to providing energy-efficient and cost-effective space conditioning for offices and other commercial and industrial buildings. The application of UFAD technology, in particular, showed strong growth in North America during the 1990s. Today, it is estimated that approximately 12 percent of new office buildings in North America feature raised floors and, of those, around 40 percent include underfloor air systems.

As KEEN had discovered, the European – and especially German – building codes define and mandate interior air movement velocities, so that air distribution is a highly engineered aspect of mechanical systems. Displacement ventilation systems have also been more commonly used there to keep air velocities in occupied spaces at less than fifty feet per minute (fpm) (0.25 meters per second). Meanwhile, North American buildings, including many designed by KEEN, still predominantly feature traditional overhead air distribution systems according to client specifications or budget constraints.

While conventional overhead systems are designed to produce a single uniform thermal and ventilation environment within each building zone, UFAD systems deliver air through smaller-sized supply outlets distributed across the floor plate. In effect, this feature gives building occupants nearby some degree of personal comfort control, depending on diffuser design and location. This design approach makes use of the open space between a structural slab and the underside of a raised floor system to deliver the conditioned air supply within the occupied zone of the space. In a typical installation, swirl diffusers are used to provide greater air mixing in the occupied zone (up to 6 feet). The reduced airflow requirements and lower static pressure of UFAD may eliminate as much as 90 percent of a standard building's ductwork.

Case Study

KEEN's North Vancouver Office

With a rapidly expanding staff in its North Vancouver office, KEEN Engineering was in desperate need of a new office. The firm wanted a flexible space that would encourage good communication and allow for frequent workgroup changes in response to new projects. A 16,000-square-foot office was selected, large enough to house all engineering staff on one floor, while a surrounding 6,000-square-foot mezzanine was designated for all non-engineering support staff. KEEN also wanted the facility to serve as a showpiece for its innovative mechanical design ideas – and it would be designed by KEEN engineers.

KEEN opted for an underfloor air approach, combined with a state-of-the-art direct digital control system, to provide both flexibility and energy efficiency. This feature solved the problem of continual staff and computer reconfiguration by making power, computer, and communications cables easily accessible. The resulting air system delivers conditioned air from beneath the walking surface through diffusers flush to the floor. Each employee has at least one diffuser in his or her workspace. Older, stale air migrates toward the ceiling due to thermal lift and is removed at the ceiling through a combination of exhaust and natural relief air vents.

Upon the completion of the office and the mechanical system in January 1992, KEEN staff observed a noticeable improvement in air quality. In addition, the UFAD system immediately saved up to 20 percent per year in energy consumption through the use of natural heat stratification and decreased airflow resistance. And this energy saving was realized even though, in 1992 alone, eighty-five tours were conducted.

High-Tech Lesson V: *UFAD systems mean both higher and lower first costs compared to overhead systems. The raised flooring itself and physically sealing the plenum at the connections between the building skin and slab is more costly. However, the elimination of overhead ductwork, downsized HVAC equipment, and lower floor-to-floor heights all significantly reduce costs as well.*

High-Tech Lesson VI: *The dynamics of underfloor air calculations are completely different from those of conventional overhead systems. The control and optimization of stratification is crucial to system design and to sizing, energy-efficient cooling operation, and comfort performance of UFAD.*

The closely-related displacement ventilation systems differ primarily in the way air is delivered to spaces. Used for cooling purposes only, DV system air is supplied at very low velocities through low side-wall diffusers near floor level and returned near ceiling level. Because supply air is delivered directly into the occupied zone, it is introduced at a temperature only slightly (5°F) below comfort conditions. Incoming supply air has very little momentum; as this cooler, heavier air enters the space, it "falls over" to become a five- or six-inch thick layer spreading across the floor, much like flowing water. The upward movement of air in the room takes advantage of the natural buoyancy of heat gain in the space, producing a vertical temperature gradient. This principle is the basis for the improved ventilation effectiveness and heat-removal efficiency associated with DV systems.

A Closer Look at UFAD Systems

The rapid growth of UFAD applications in North America is due, in large part, to the increase in green buildings. Today, the U.S. Government Services Administration and many state and local municipalities are requiring LEED® certification for all buildings. Underfloor air systems were originally introduced in the 1950s for spaces experiencing high heat loads, such as computer rooms, and later became more widespread in European offices by the mid-1970s.

KEEN's earliest design experience with raised flooring came in 1978, although no underfloor air systems were involved. By the mid-1980s, design was underway on the new headquarters for BC Hydro, British Columbia's major electric utility – a 500,000-square-foot project that would represent the first and largest commercial use of an underfloor air system in North America. Simultaneously, KEEN engineers would design and install a UFAD system for their own offices in North Vancouver. And then came the breakthrough system designs for the $100 million multi-building Library Square complex. By 1998, KEEN was clearly and unquestionably recognized as the North American engineering leader in underfloor air design.

Challenges/Opportunities

UFAD offers a broad range of important advantages over conventional overhead systems in delivering conditioned air to localized diffusers in the occupied zone of a building. Modularity and flexibility are two of the most important advantages. Installation of raised flooring, which forms an underfloor plenum, provides greater access to building services and makes workstations easier to re-configure. As a result, "churn" can be less expensive. Many building owners are also finding that by integrating their flooring, HVAC, power and data lines, they can build out tenant space faster and shorten time to occupancy.

The performance advantages of UFAD systems arise from the fact that conditioned air is delivered at or near floor level directly into a building's occupied zones, and returned at or near ceiling level. In the cooling mode, these systems also deliver better heat removal and lower operating costs. In addition, because the system is typically one large pressurized plenum, an entire open plan area maintains a consistent climate. This approach allows occupants to control fresh air directly, improving room ventilation and indoor air quality. Well-designed UFAD systems can also reduce a building's HVAC energy through the use of lower supply-air temperatures, lower-horsepower fans, extended economizer cooling, and improved cooling-cycle efficiency. This energy reduction, of course, depends on building design and location.

Best Practices

Compared to the early UFAD applications – where overhead products were often simply adapted under the floor – a number of manufacturers have developed products specifically tailored to UFAD systems. When using the underfloor plenum as an air delivery passageway, designers should consider one of three basic air-distribution methods: pressurized plenum, zero-pressure plenum, and ducted air supply. Today's raised floor systems typically consist of 24-inch by 24-inch composite steel and concrete floor panels fastened to supporting pedestals. The pedestals, with integrated leveling devices, are bonded to the slab with adhesive. Installations often end up as hybrid solutions, featuring some ductwork depending on the size of the floor plate, the number of shafts and other system design features.

Design Considerations

North American traditionalism in mechanical design has caused many engineers to avoid new approaches and systems such as underfloor air that are unfamiliar – where, as McCarry said, "there were few if any North American examples available to go over and kick the tires." Until recently, manufacturers and standards organizations had not provided the necessary tools to properly assess the performance and economics of underfloor air distribution systems.

A somewhat common problem for designers of new UFAD projects is to assume that load calculation procedures and programs used for conventional overhead air design will also work for UFAD. Many engineers still inaccurately assume that UFAD systems, with warmer supply air temperatures (63°F to 65°F), will automatically require higher cooling air quantities in comparison to overhead mixing systems (55°F supply-air temperature) to maintain equivalent comfort for the same space loads.

Initially, UFAD designs called for floor outlets to be combined with adjustable desk outlets. Now, floor outlets predominate. As a rule of thumb, design requires approximately 100-square-feet of floor space and one to two floor outlets of 8-inch diameter per person, or three to six floor outlets of 6-inch diameter. The supply air temperature should be at least 64°F (18°C), with minimum distances between a fixed workplace and air outlet of 2 feet.

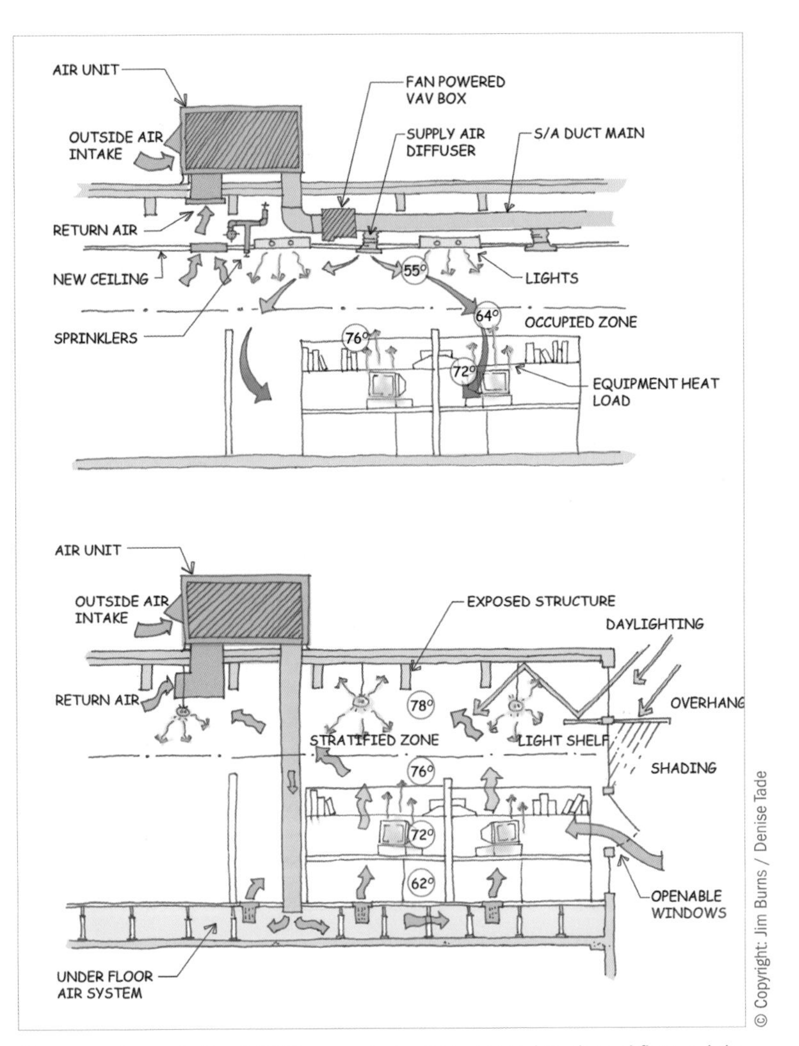

Air overhead or underfloor? KEEN concept sketches are used to show airflow and the performance differences of each system within the occupied zones of a building.

A Closer Look at DV Systems

Although it relies on many of the same principles as UFAD systems for cooling performance, displacement ventilation is still not commonly used in North America. This technique makes effective use of the zones, placing fresh air where it is needed near occupants and using the natural stratification of air temperatures to its advantage. With displacement ventilation, fresh cool air is supplied at floor level in the occupied zone and warm contaminants rise to the ceiling where the contaminated air is exhausted from the space.

DV design techniques first gained popularity in industrial applications during the 1970s, mostly in Scandinavian countries, and became more widely used for commercial structures in the 1980s. On one of his research trips to Europe, Blair McCarry found DV in use on a broad range of applications including offices, schools, banks, and theaters. He discovered that exterior loads of buildings are controlled more stringently in Europe than in North America, allowing for a much broader range of usage possibilities.

KEEN's first serious displacement ventilation project to provide user comfort was for the new Morris J. Wosk Centre for Dialogue, at Simon Fraser University in Vancouver, in late 1997. As part of this 42,000-square-foot, five-story renovation, KEEN specified a DV system for its Dialogue Hall, a unique circular seating space where supply air was delivered at the floor level for conference attendees.

Challenges/Opportunities

Displacement ventilation has two main advantages over traditional overhead air systems. First, it can improve indoor air quality throughout occupied spaces by effectively removing contaminants associated with heat sources – such as computers, people and equipment – to the upper part of a room where exhausts are located. Second, a DV system tends to use energy more efficiently as it removes exhaust air from the room at a higher temperature than that in the occupied zone; therefore, less cooling is needed for a given temperature in the occupied spaces.

In addition, in a number of completed projects KEEN has found that the combination of DV and radiant heating or cooling, for example, may actually reduce the operating costs of the mechanical system. As a result, those cost savings may be transferred over to another part of the project, such as a better building envelope that, in turn, allows designers to downsize the building's boiler or chiller.

Best Practices

The primary driving force for displacement ventilation is to achieve relatively low cooling demand. Practitioners agree that the optimal DV applications include: areas with high ceilings where routing for low level air supply is available; spaces in which cooling loads have been minimized, so heating and cooling can be supplemented using radiant systems and ensure good air quality and temperature control; and in spaces where the air system is not the primary source of heating or cooling.

> It's always been important for me to investigate how underfloor air or displacement ventilation works – seeing a lot of projects, kicking the tires, noting the climate and comfort differences. But this one time I realized I just didn't have it yet. I could talk it, sell it and diagram it, but I didn't understand the engineering behind it well enough. So I set up another trip.
>
> – Blair McCarry, KEEN

To date, KEEN has applied DV systems with wall grilles to a number of school classrooms and lecture theaters, also using the technique in airports for lounge and ticket areas. It also has excellent performance in auditoriums and theaters with tiered seating; in this setting, displacement air is tied into the floor or seats, where gentle, draft-free air is supplied at a low level in seating areas and allowed to stratify up to high ceilings.

Design Considerations

Displacement ventilation can provide a comfortable indoor environment at a high cooling load through careful design. Yet, it is important to note that most North American cities typically have higher temperatures in summer than those in Scandinavian cities, and U.S. offices may have more lighting and equipment that produce more heat. Therefore, cooling loads could be higher.

DV system configurations include raised access flooring using floor grilles or perimeter baseboard style air diffusers – both effective for room air diffusion and ventilation at low velocities. Designs may also include freestanding desks or even two-foot round grilles that rise up approximately three feet off the ground, like a cylinder, with perforated air grilles – an approach used in airports. An additional design, typically for high-traffic industrial or commercial settings, features a duct descending to about ten feet off the floor; the cool air is then allowed to pour out like a waterfall into occupied spaces. KEEN has also installed wall grilles above doors in high-traffic areas where they perform best.

During design development, it may also be useful to perform computational fluid dynamics modeling to understand the potential performance of a DV system. Most systems are designed for a maximum outlet air velocity of 40 fpm (0.20 m/s) at a minimum supply air temperature of 17°C (63°F).

Pacific Press Kennedy Heights Printing Plant at night. Displacement ventilation serves the Press Hall featured here.

Further UFAD and DV Resources

ASHRAE's "Underfloor Air Distribution (UFAD) Design Guide" (2003) describes some of the advantages of UFAD systems – improved thermal comfort, improved ventilation efficiency and reduced life-cycle building costs – and provides assistance in the design of high performance UFAD systems.

ASHRAE's "System Performance Evaluation and Design Guidelines for Displacement Ventilation" (2003) is a ten step guideline for use by designers in designing displacement ventilation systems – the result of an ASHRAE research project by Quingyna Chen, Ph.D. and Leon Glickman, Ph.D.

"Displacement Ventilation in Non-Industrial Premises" (2001) is a comprehensive design manual covering displacement in commercial applications. Published by the Federation of European Heating and Air-Conditioning Associations (REHVA), it is sold in the United States by ASHRAE.

DV and UFAD modules are in development for EnergyPlus to facilitate more accurate energy modeling of these alternative systems.

Case Study

Pacific Press

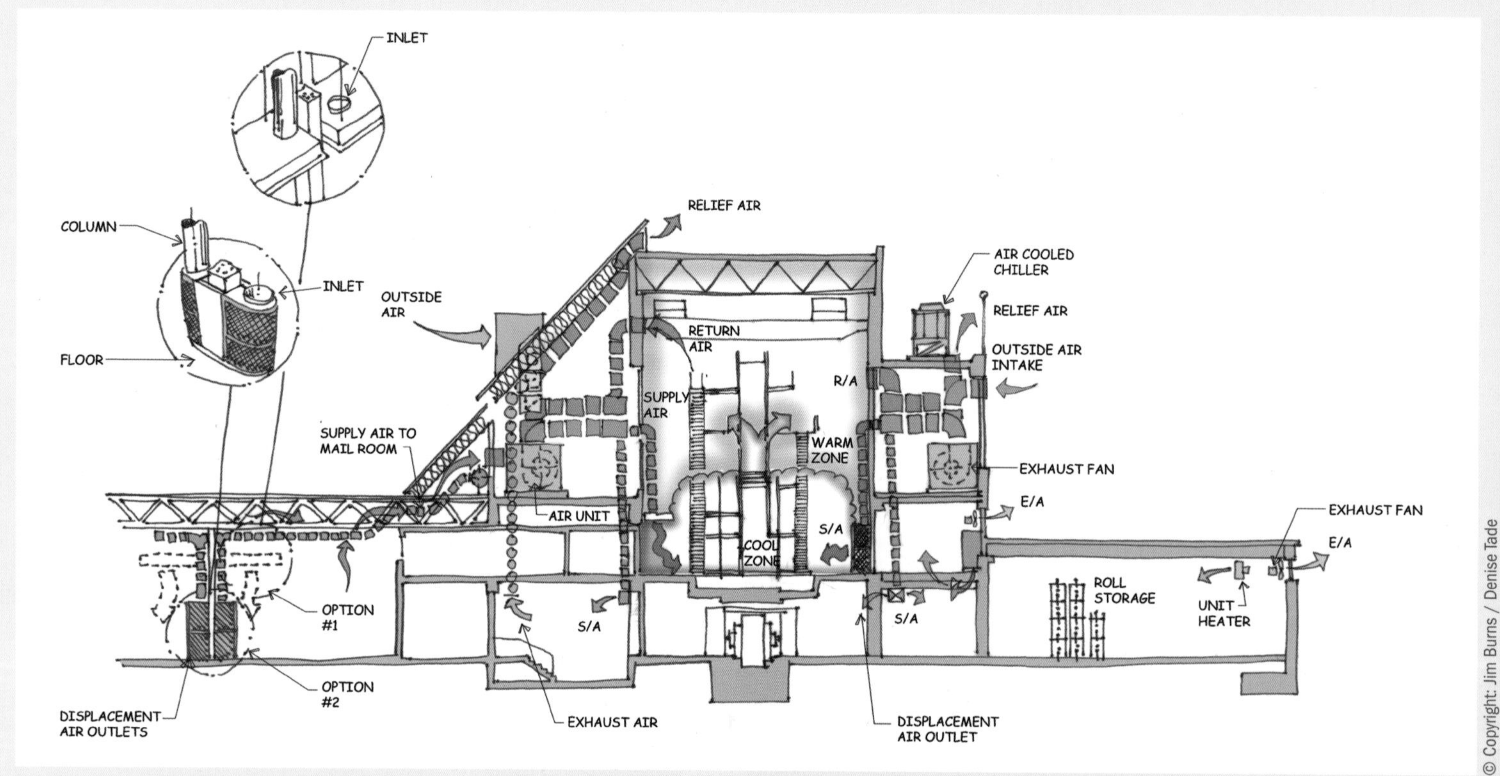

For the new Pacific Press facility design by Stantec, KEEN Engineering designed an innovative displacement ventilation scheme that introduces cool air at floor level and efficiently funnels the heat from its large printing presses to the return air system.

Pacific Press, owners and printers of the *Vancouver Sun and Province*, were in need of a new production facility to support their expanding business. With the purchase of several large presses from Europe, due to arrive in December 1996, the fast-track design and construction of a weather-tight press hall structure had to be completed in just seven months. For KEEN, the biggest challenge would be heat removal from the building – since newspaper production is an energy-intensive process and traditional methods of removing heat would inevitably add to the client's already-high energy costs.

Blair McCarry of KEEN went on a fact-finding trip to Europe to investigate the possibility of using displacement ventilation within the press hall – an approach never used in a facility like this before in North America. He met with industrial DV (Direct Ventilation) manufacturers in Germany and local engineers who had experience with thermal stratification in comparable projects.

As designed, the system introduces cooled air into the press hall and mailroom at floor level, two areas which account for more than 80 percent of the building's cooling load. The 60-foot-high press hall contains eight large presses, with aisles in between for forklift trucks to transport paper back and forth. In each aisle, three-foot-diameter metal ducts descend to approximately ten feet off the floor, capped with displacement grilles; as the cool supply air wafts down onto printing crews, it then moves horizontally in the space. Heat generated by the printing presses causes the warm air to rise naturally to a high level, where it is funneled into the return air system.

As a result of this innovative DV approach, the natural heat rising from the press equipment has allowed the client to cut energy costs significantly. In addition, the system has led to an estimated savings of 250 – 300 tons of refrigeration and a reduction in the size of cooling equipment by one-half.

High-Tech Lesson VII: *If displacement ventilation is not powerful enough to meet the full cooling requirements, it could be combined with a radiant system – which may actually result in a cost reduction overall as the total air system is downsized dramatically.*

High-Tech Lesson VIII: *There are essentially two grades of displacement ventilation: one for industrial areas and one for offices and other public spaces. The acoustically quiet air is ideal for performance halls. It may also be perfect for library environments, by creating layers of temperature in a space with very low velocities of cool air to help maintain the physical condition of books.*

Part Three

The Power of Touch: Radiant Heating and Cooling Systems

Radiant heating and cooling, including the use of in-slab radiant systems and radiant chilled ceiling panels, can be very effective techniques to maintain a near-constant space comfort condition, even in lobbies and other transient spaces where room temperatures may vary widely due to fluctuating solar loads and other environmental factors. In effect, radiant systems operate largely on the principle of thermal radiation – heating and cooling that occurs between surfaces or between a source and a surface. Attaining adequate thermal comfort addresses the mean-radiant temperature of a space and the instantaneous heat transfer between the surface of the radiant slab and other sources of warmth or coolness nearby. This technique essentially offers some control of radiant heat loss or gain once a building occupant enters a room. Similarly, where increased solar loads coming in through windows may heat up interior objects in an office space or atrium, radiant slabs can offset that radiant heat exchange from the glass. Based on these unique characteristics, radiant systems can be effective for both spot heating and space heating or cooling requirements for an entire building.

Although a few radiant installations in North America date back as early as the 1960s, most have performed inadequately because of either misunderstood methods or inadequate equipment. Today, a growing number of large-scale radiant slab building systems are in use around the world. And radiant system technologies have vastly improved – thanks, in large part, to advancements through projects in Switzerland and other parts of Europe over the past two decades.

KEEN's first real exposure to radiant heating and cooling came with the design and construction of Electronic Arts' new 200,000-square-foot Canadian headquarters in Burnaby, British Columbia. The lead designer on the project, Blair McCarry, used a few basic calculations and intuitive strategies to create the successful radiant system.

He was intrigued to learn still more about these relatively unknown, multi-floor, slab-on-grade techniques. However, there were few radiant heating and cooling resources available on this side of the Atlantic. So, on repeated "research jaunts" to Europe, McCarry was able to tour several buildings to learn directly from their designers about the unique dynamics and temperature variations of these systems, including the use of radiant chilled ceiling panels. One visit included the Dow Building, headquarters for Dow Europe, outside Zurich – home to one of the largest and earliest examples of in-slab cooling in the world.

A Closer Look at Radiant Heating and Cooling

In Switzerland, McCarry learned the critical importance of design temperature to effectively condition building spaces using radiant systems. Both McCarry and KEEN learned how very different European comfort standards and design styles were from those in North America:

> For example, going up to 80°F (26°C) for an air-conditioned building on a hot day is acceptable in Europe. A lot of the performance of slabs involves the difference between slab temperature and room temperature. So, if you're trying to keep your room at 72°F (22°C) and your slab is at a certain temperature, you have half the capacity compared to a room at 80°F.

This Swiss approach over the last fifteen years has led to radiant system designs that feature plastic tubing encased in a 6- or 8-inch concrete structural slab, enabling the building mass to work with the HVAC system as an energy storage reservoir. This in-slab tubing – a serpentine, continuous loop of 5/8-inch diameter PEX plastic piping – then distributes warm or cool water from the central boiler or chiller plants through the exposed concrete slab to control its temperature. Water can hold three thousand times as much thermal energy as air. As it flows through the piping, the water immediately begins to cool or warm the concrete mass, with the resulting temperature eventually making its way to the surface. This delay, or longer response time, compared to conventional heating devices, can pose a real challenge in controlling room temperatures from season to season. McCarry has found that in European buildings, for example, the slab is warmed or cooled only at night and then allowed to "freewheel" during the daytime. However, this approach represents a broader temperature comfort range than is generally accepted in North America; it also relies on major solar control for windows not commonly found here.

Challenges/Opportunities

After tracking the performance of those ground-breaking Swiss buildings for more than a decade, KEEN has determined that radiant office spaces use significantly less energy than conventionally heated and cooled buildings; indoor environment conditions are better; and they typically cost less to build. In addition, radiant floor heating is capable of decreasing space-conditioning loads and leveling peak loads, and the piping configuration provides increased control over thermal mass.

Equally important, these dynamic slabs couple extremely well with other low-energy heating and cooling sources as a part of an integrated approach. This integrated approach includes displacement ventilation as well as condensing boilers with low return water temperatures. Another strong tie-in is with ground-source heat pumps. When used in combination with a well-insulated building envelope, lower heating water temperatures and higher chilled water temperatures, heat pump systems use less energy more efficiently to perform the same tasks.

The primary concern with radiant heating and cooling is ambient relative humidity, particularly in areas along the East Coast, Midwest and southeastern U.S. that experience hot, humid climates. The performance of chilled ceilings and chilled slabs, for example, will be limited by the dew point of a building's indoor air. Therefore, condensation avoidance is very important – through the use of lower chilled ceiling temperatures in drier climates or dehumidification equipment in the supply air system. As a general rule, lower radiant floor cooling temperatures of approximately 60°F (16°C) will prevent any condensation from forming on floors.

In-slab radiant heating and cooling – an approach KEEN has adopted and further refined from European examples, relies on water flowing through a continuous loop of plastic tubing to warm or cool an exposed concrete mass.

Best Practices

Radiant systems have been applied successfully to many different types of buildings – with a growing number appearing in North America. By controlling surrounding surface temperatures, radiant floors and ceilings can handle the majority of occupant comfort requirements.

In temperate climates, the slab needs only to remain at a roughly constant 72°F all year round to provide adequate cooling. This constancy means the temperature may rise up to between 75°F and 77°F in the summer and drop down to 67°F or 68°F in the winter. If necessary, the slab temperature control scheme may be adjusted to restrict that range more tightly between seasons.

In more severe winter climates, supplemental perimeter heating may be needed, depending on the extent and type of glazing and how cold it gets in winter. In fact, high-performance buildings in many areas of Switzerland and Germany, where the winter design temperature is 5°F to 14°F, do not require auxiliary perimeter heat.

KEEN's own experience with radiant systems has shown that even in more extreme climates, properly-designed and applied slab temperatures can vary by just 18°F. That is the case in British Columbia's interior, where outside temperatures range widely from -31°F to 95°F during the year. In 2003, KEEN completed a system near Quesnel for North Cariboo Community College (NCCC). As designed, the school now relies on a combination of in-slab radiant flooring and displacement ventilation for heating and cooling; there, surface slab temperatures range from 66°F to 82°F, representing just a sixteen degree variation to accommodate a one hundred-twenty degree swing overall in air temperature.

At NCCC, the radiant system takes advantage of a ground-source heat pump for lower supply temperatures during heating. It supplies higher temperature during cooling. In summer, water pumped from a well passes through a heat exchanger and is then fed through the in-slab piping to radiate cooling to occupants; an air handling unit draws in fresh outdoor air and cools it to approximately 68°F; the air is then dispersed to various zones via variable air volume units. In winter, the well water passes through the heat exchanger and boiler, branching off through the in-slab piping to radiate heat to occupants.

Design Considerations

The key to a successful radiant system begins with the building envelope itself. A high-performance envelope can minimize exterior loads, thus enabling the thermal storage of a radiant slab to effectively manage the loads of meeting rooms and offices. Maintaining that constant temperature becomes especially important in transient areas such as atria, where sunlight can heat up the floor and radiate off into the space, heating it up as well. Instead, KEEN tries to maintain that floor at a temperature of 68°F to 70°F, so the radiant slab becomes a powerful solution for heat absorption.

It is also important for design teams to minimize thermal fluctuations in a given space in keeping with the slab's slow response to heat losses and heat gains. As a result of the in-slab piping, it is not necessary to generate cooling around the clock; instead, the slab may be cooled in off-peak hours, which in turn allows designers to reduce the size of the chiller. In addition, there are alternatives to concrete slab systems for radiant heating or cooling. Metal radiant heating/cooling panels with copper tubing attached to the back have been used for years for heating in North America and are also used for cooling in Australia. These metal panels can integrate into a number of ceiling systems, but are more expensive than in slab piping.

Whether incorporating metal or concrete, radiant systems are designed simply to generate a cooling or warming effect, using large surface areas and resulting in a stable indoor climate. The principle here is that air is not heated or cooled to condition these spaces. Therefore, radiant systems often eliminate the need for fans, reduce energy consumption, and free up displacement ventilation or some other energy-efficient HVAC system to concentrate on meeting ventilation requirements.

Case Study

ICICS Building Expansion

When the University of British Columbia, Vancouver, decided to expand its ICICS (Institute for Computer Information & Cognitive Studies) Building, campus planners also wanted to meet requirements of LEED® Silver through a variety of green strategies aimed at reducing energy use. As a result, the design team – which included KEEN – recommended a high-performance building envelope and window glazing, natural ventilation, underfloor air systems – and a radiant chilled/heating slab system.

Completed in December 2004, the new facilities include a six-story, 107,600 square-foot addition to the existing ICICS Building, and the separate 16,150-square-foot Lecture Pavilion. Together, the two components provide additional office and research space, lecture theaters and classrooms for the Institute. Three levels feature access flooring and three levels have concrete floors with embedded radiant systems that use the building's thermal mass and inertia to normalize space temperatures. This method decouples the cooling energy of the ventilation and chilled slab systems to reduce the air handling system size by nearly 75 percent.

According to McCarry, the chilled slab system represented the most significant sustainability challenge. The system called for an exposed concrete slab overhead so that most of the cooling energy could radiate down from the ceiling. The exposed structure required extensive design collaboration among all consultants to coordinate lighting, structural and architectural features with the exposed mechanical ductwork and piping systems. Controlling the chilled slab was also mechanically challenging. It requires nine to twelve hours of lead time to change temperature because of its thermal mass; therefore, controls were included to enable fine-tuning of the system after the initial year or two of temperature trend logging.

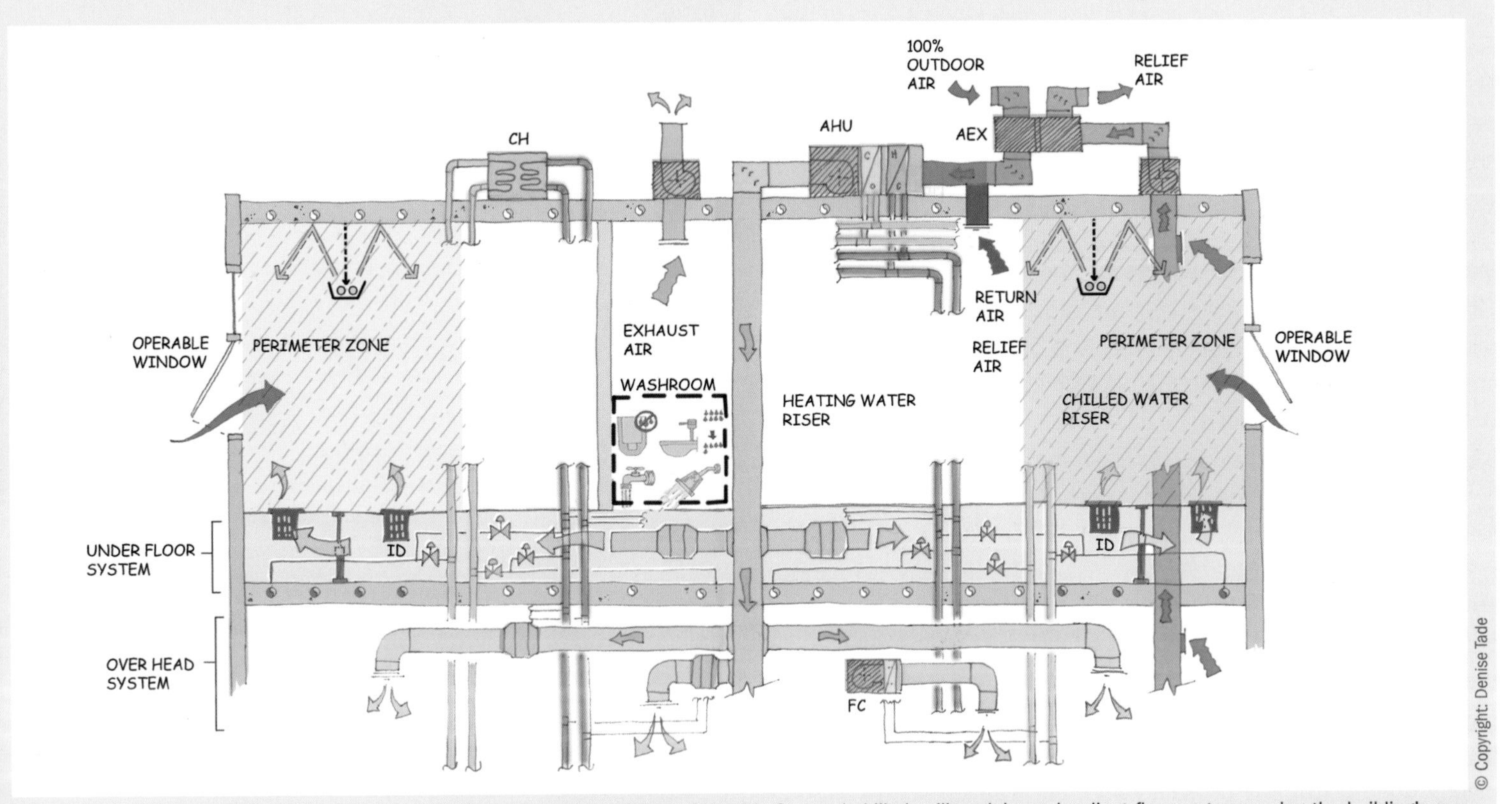

KEEN's concept for the ICICS Building (Vancouver) at the University of British Columbia featured chilled ceiling slabs and radiant floor systems, using the building's own thermal mass to normalize temperatures within occupied spaces.

High-Tech Lesson VII: *Instead of focusing solely on the high peak thermal loads and short response times typically created by low performance envelopes, radiant calculations and sizing should factor in the resultant temperature (simple average of the mean radiant temperature and the ambient air temperature) of a space for good thermal comfort.*

High-Tech Lesson VIII: *Always pay attention to acoustics and other environmental factors when choosing a system. For example, a concrete floor, concrete ceiling, drywall and glass walls can be a very loud space, with no 'white' noise produced by radiant systems to mask sound. Do solutions match functions?*

Through KEEN's own design learning curve with radiant systems, its engineers have relied primarily on two tools. The first tool involves TAS (Thermal Analysis Software) to simulate the performance of thermal mass in a given space once the slab's surface temperature is known. However, TAS is not able to calculate the water-to-concrete heat transfer. So the KEEN team uses an iterative process of feeding data back and forth between TAS and finite element analysis to model room temperatures and assess how the internal dynamics of the slab is affected by different water temperatures and flow rates.

More recently, they have adopted a newer simulation program, TRNSYSlite, to model user comfort requirements based on the relationship of a ventilation system to a proposed chilled slab on one project and to determine the power output necessary for in-slab cooling on another project.

ICICS Tower Exterior.

Conclusion

What high-tech solutions are next for KEEN? Already, its design engineers are exploring new alternatives for managing site water, expanding greywater use, minimizing overhead ductwork and air systems, and reducing floor-to-floor heights. Breakthrough projects are opening up new opportunities for low-energy cooling and innovative biological waste treatment systems. At KEEN, "high-tech" means staying focused on technology, but also on subtle, precise design schemes – and always on taking a First Principles approach to engineering.

For KEEN these new realities – these design truths – continue to emerge and evolve: that the capital cost of these sustainable solutions are often equal to or less than a conventional building approach; that the synergies of combining low-tech and high-tech strategies can achieve higher levels of thermal comfort and energy efficiency than ever before; and that KEEN is actively helping to move North America's MEP industry to a new "norm" of sustainability – one project at a time.

Further Radiant Heating/Cooling Resources

"Radiant Floor Heating in Theory and Practice," Bjarne W. Olesen, *ASHRAE Journal*, July 2002.

"Radiant Heating and Cooling, Displacement Ventilation with Heat Recovery and Storm Water Cooling: An Environmentally Responsible HVAC System", Stephen C. Carpenter, PE and John P. Kokko, PE, *ASHRAE Transactions*, 1998.

Section Three

The Projects

Moving and turning above us, around us, within us is the invisible element that first animated the planet – the breath of life.

– David Suzuki, *The Sacred Balance*

York University Computer Science Building

The Future Has Arrived

Toronto, Ontario

Autonomous robotics, computer vision, theoretical models of computation, artificial intelligence, human and computer interaction–all are part of the future of computing. And the future has already arrived at Canada's York University.

Advanced research interests like these have made York's Department of Computer Science and Engineering one of the leading academic and research programs in Canada today. In addition to its dynamic research environment for faculty, staff and graduate students, the department is home to over fifteen hundred computer science majors.

It is fitting, then, that Computer Science and Engineering is located

The first cold-weather climate green building in North America, York's Computer Science Building features exterior shading and "blank-out" louvres above the main entrance (south elevation).

Primary Team Members

Owner:
York University

Architects:
Busby + Associates, Vancouver, BC
Project Team: P. Busby (Partner-in-Charge), V. Gillies, M. McColl (Associate-in-Charge), A. Waugh, S. Ockwell, J. South
ArchitectsAlliance, Toronto, ON
Project Team: A. DiCastri (Partner-in-Charge), W. Bettio (Project Architect), B. Zee, M. Lukasik

Engineers:
Mechanical: Keen Engineering, Toronto, ON – M. Godawa (Engineer of Record) K. Hydes (Partner in charge), W. Cheng (Project Engineer)
Structural: Yolles Partnership Ltd., Toronto, ON
Electrical: Carinci Burt Rogers Inc., Toronto, ON

Consultants:
Landscape Architecture: John Lloyd & Associates, Toronto, ON
Environmental Sustainability: Berkebile Nelson Immenschuh McDowell (BNIM) Architects, Kansas City, MO
Cost Consultant: Hanscomb Consultants, Toronto, ON

General Contractor:
Ellis-Don Construction Ltd., Mississauga, ON

in one of the most advanced cold-climate green buildings in North America. In 2002, *World Architecture Magazine* chose the York University Computer Science Building in Toronto, Canada as one of the greenest buildings in the world.

Located at the northern edge of metropolitan Toronto, York is one of the largest universities in the country. Sustained growth over the past twenty years, due, in part, to surging enrollment in computer science programs, led the University in 1998 to commission Busby + Associates Architects, in joint venture with Van Nostrand DiCastri Architects, to design a new dedicated Computer Science facility.

Environmental Objectives

From the outset of project planning, the University's goal was that this construction would be the first green institutional building project in Ontario. Peter Struk, former head of facilities at York, was instrumental in initiating a sustainability mandate for the University which, in turn, offered the project team an opportunity to prove the design approach could work in a "cold weather" climate. Since then, all buildings on the York campus must meet environmental sustainability objectives.

The project kicked off with an all-day design charrette at York, led by Bob Berkebile of BNIM Architects in Kansas City, Missouri. Participants included the users and the architectural and engineering team. Their charge: to define sustainability goals and identify features that would lead to good indoor air quality and user satisfaction while not exceeding the capital budget. In fact, Struk and the University challenged the team to design a green building that would not cost any more than standard construction and yet achieve energy performance resulting in lower operating costs.

Design Overview

The University directed the design team to create a warm, open, welcoming facility with as much natural light as possible through the use of skylights and translucent glass substituting for drywall partitions

The building's "Tree Atrium" near the main entrance promotes fresh air and daylight throughout the three-story space and into adjacent offices.

and yet be simple and flexible enough to accommodate unpredictable future technology. The Computer Science Building opens up with a large glass elevation, a welcoming entry and open circulation space.

The 115,000 square-foot, four-level building is organized around three functional components: a two-level block with lecture theaters; computer classrooms and research and teaching laboratories placed on the north side to capture efficiencies in energy consumption; and

As designed by Busby + Associates with Architects Alliance, the sawtooth design (west elevation) excludes direct sunlight to minimize heat gain – enabling offices to capture the more controllable southern light.

> The question was, could we deliver the same levels of performance and energy savings in a more extreme climate. We had faith that our ideas were going to come through, and we had European examples to back us up. We wanted to convince them – and they were ready to be convinced.
>
> – Peter Busby

Project Program and Stats

Location:	York University Campus, North York (Toronto, Ontario)
Building Type:	Educational Facility, New Construction, Single Structure
Size:	Approx. 115,000 sq. ft.
Stories:	Three-story building, with a fourth below grade
Building Features:	Dedicated research space with restricted access; computer labs; three lecture theaters (one with 500 seats and two with approx. 200 seats each); graduate offices
Site Issues:	Greenfield site, placed in between other campus buildings that needed to stay operational
Completion Date:	August 2001
Certification:	CBIP – Commercial Building Incentive Program
Cost:	$17 million
Awards:	· Consulting Engineers of Canada Award of Excellence, 2002 · "The Best Green Buildings", *World Architecture Magazine*, September 2002 · Architectural Institute of British Columbia "Lieutenant Governor of BC Medal"

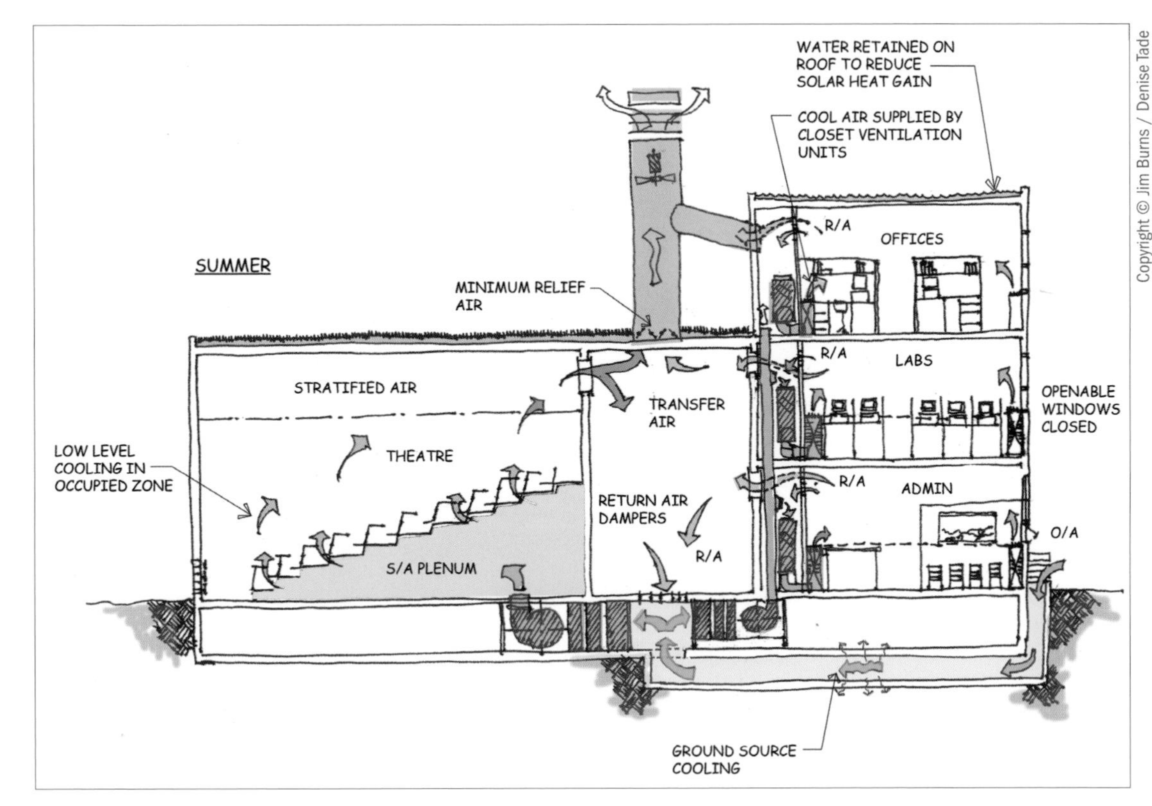

KEEN's concept sketch illustrates expected air movement within the building in summer: from the subsurface airflow plenum through the lecture theaters and air handling units to other occupied spaces, and finally skylights and solar chimneys.

communications systems and faculty and research offices located in a flexible wing on the west side that parallels the main circulation space. All are connected through two large atria.

The east and west sides of the building feature a sawtooth design, with the glazing angled to the southeast and northwest, respectively, to exclude direct sunlight while admitting as much natural light as possible. This orientation minimizes heat gain from the harsh eastern or western sun, while allowing office windows to capture more "controllable" southern light. On the south side, the building features exterior shading and "blank-out" louvres, which allow a speaker to control the amount of light coming into the lecture theater with moveable shades for presentations and other activities.

Also important to the design and performance of the building are two atria: the Tree Atrium (with bamboo trees) near the main entrance and the Bar Atrium (the research wing). Given the relatively deep floor plate, both areas are integral to enhancing natural ventilation and providing daylighting within the Computer Science Building. This high-tech building uses low-tech solutions to exceed the ASHRAE 90.1 standard for energy efficiency by 40 percent.

MEP Design Process

The building's energy performance is achieved through several integrated techniques. The primary design element is natural ventilation – a real challenge in an internally load-dominated building in the Ontario climate, with four seasons and a winter design temperature of - 4° F. The exposed structure (12-inch thick slabs, all concrete) also provides thermal mass to smooth heating and cooling load peaks. In addition, external shading on the south exposure and saw-tooth walls on the east and west sides neutralize unwanted heat gain, thereby further reducing the cooling load. Underfloor air delivery in the lecture theatres promote the stratification of heat that delivers air to the occupant breathing zone and thereby improves indoor air quality and reduces cooling loads.

Using a mixed-mode design approach, KEEN's engineers relied on two separate strategies: Summer/Winter and Spring/Fall. During the milder Spring and Fall shoulder seasons, the building opens up as fresh air flows through operable windows into classrooms and office spaces. The air then relieves into the corridors, which are interconnected to the atria. The air transferred to the atria is relieved through motorized high-level openings. Wind direction sensors control these high-level openings in order to eliminate potential downdrafts which would undermine the system. To extend the effectiveness of natural ventilation during hotter summer months, dual-duty smoke exhaust fans can be run at 50 percent speed with a variable speed drive to help pull air through the building.

In Summer/Winter mode, the space is essentially "buttoned-up" and mechanical systems are used to provide cooling/heating to the space. One air handler delivers fresh, tempered air to the atria space. A second air-handling unit serves the two basement lecture theatres and the large theatre via an underfloor supply system. A modular fan coil design is used for the perimeter and internal spaces. To serve peak summer and winter loads, these fan coils take air from the atria, condition it and deliver it into the space. The air is then partially relieved back into the atria, which serves as a mixing plenum for the fresh air delivered by the air-handling unit and the return air from the classrooms, offices and theaters. The remainder of the air delivered by the fan coils is exhausted through exhaust fans on the roof. The atria are therefore indirectly conditioned by transferred air from the occupied space.

Given the design's high thermal mass and limited use of ducts, the sheet metal and mechanical contracts were very small – 13 percent of the total cost of construction, as compared to traditional mechanical costs that typically comprise 20 percent to 25 percent of the total. By reducing the plant size and ductwork, KEEN was able to transfer significant mechanical cost savings to architectural and other project elements, realizing substantial value gains.

Indoor Environmental Quality

The Computer Science Building relies on a passive ventilation system built under the structure. Intake airways on the north

Copyright © Steven Evans, Photographer; architectsAlliance, in joint venture with Busby + Associates

Copyright © Jim Burns

side are used to pull large amounts of fresh air into the building. A concrete airflow plenum below the building then allows air to be directed where needed and leads to the three lecture theaters. As a result, occupants in the theaters receive 100 percent outside air via underfloor air vents beneath the seats. To maintain good indoor air quality, CO_2 sensors monitor how much air is necessary based on room occupancy. Controllable VAV (variable air volume) air handling units draw stale air in through vents, then filter it, condition it, and then distribute it through the concrete plenum to other parts of the building. Strategically-placed skylights and chimneys also enable the passive system

Left: KEEN's recommendation to relax the thermal requirements within the lobby and other transient areas was critical to the building's overall energy performance.

Above: Sawtooth design detail on the building's eastern elevation.

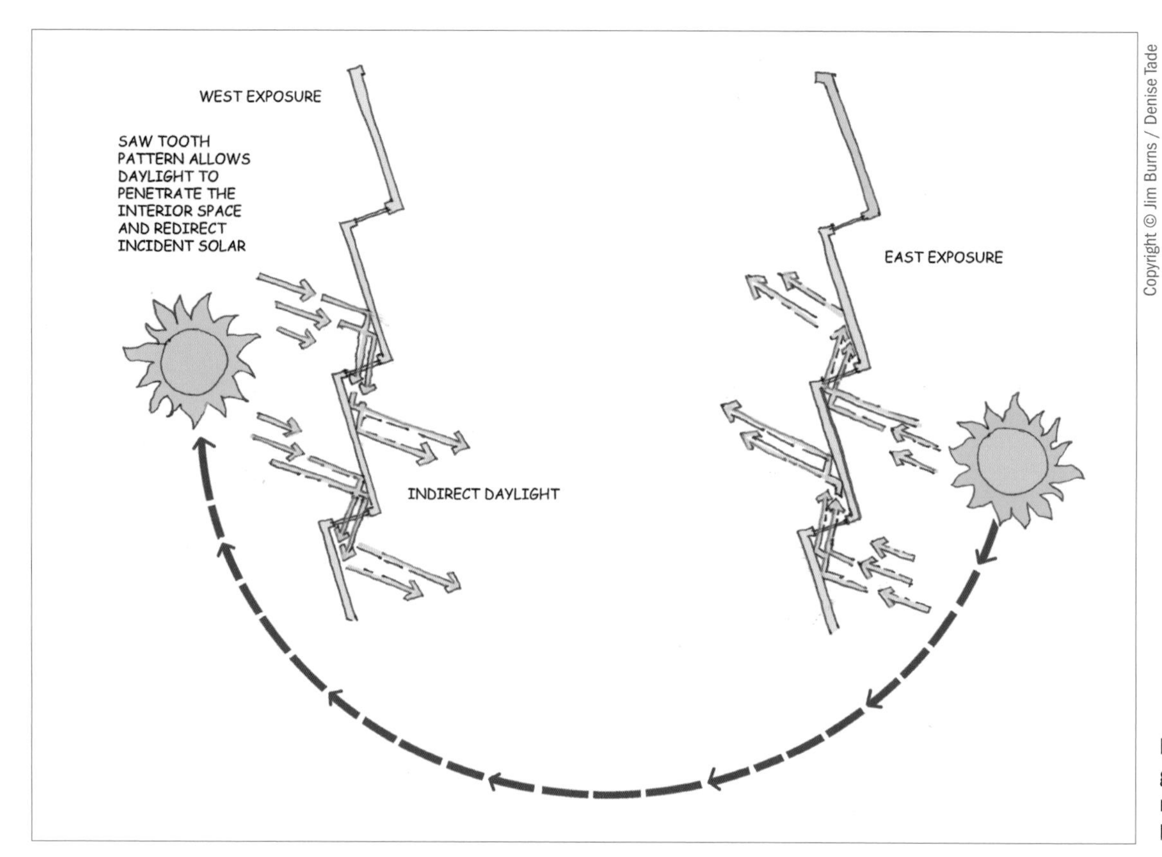

KEEN concept sketch illustrates how glazing oriented to the southeast and northwest minimizes heat gain within building spaces.

to move air efficiently and quickly where needed, using a "stack effect" that draws air out through the roof and replaces it with fresh air from the plenum.

The motorized skylights and clerestory windows on the third floor are controlled by computerized "meteorological stations" on top of the atria. These stations monitor wind, rain and temperature and are linked to the building management system to open and close windows and vents quickly according to changes in outdoor weather conditions.

The building's second air handling unit supplies air for the two atrium ventilation pools. For the "Tree Atrium," air is brought in low and allowed to filter up to a series of adjacent offices above. Offices with outside windows may be opened if conditions are right; offices facing onto the atrium also have operable windows, providing occupants with 100 percent outside air that is heated, cooled and treated. In addition, fresh air in both atriums is moved into fan coil units located in small closets around the perimeter of the building. The air is then treated and/or cooled if necessary and fed up to and exchanged through architectural cabinetry. This use of perimeter millwork as a work surface, air way and electrical chase – in combination with natural ventilation – greatly reduced the need for ductwork in the building and provides a great example of design integration.

Water and Waste Design

Stormwater management is accomplished with a sod roof consisting of an eight-inch layer of soil under grass and a wildflower garden, all covering a concrete roof. The roof over the main lecture theater is terraced to support the soil. Rooftop soil and grass naturally retain the majority of the rainwater. The design also features a series of six-foot diameter concrete rainwater catchments. Through the use of flow control roof drains, excess rainwater runoff may be re-used or discharged slowly to the stormwater management system.

The building also includes various low-flow plumbing fixtures to further reduce water consumption.

Heating and Cooling

Essential to overall building efficiency is the reduction of heating and cooling loads through a highly energy-efficient envelope. The Computer Science Building has R-20 walls, an R-30 roof, highly efficient windows with thermally broken frames (U=0.32, SC=0.45), and no thermal bridging. The internal exposed concrete structure provides thermal mass, while the sod roof reduces heat losses in winter and solar gain in summer.

To reduce energy consumption and retain the building's passive design, it was determined that a broader range of indoor air temperatures and humidity must be tolerated in the space. During summer, temperatures in the lecture theaters can reach 78°F in the stratified zone; but in the occupied zone, the temperature is 75°F. In winter, the atria is maintained at roughly 65°F to 68°F, while the classrooms and offices have a setpoint of 70°F to 72°F. This stepped approach to temperature setpoints is effective, because it graduates the occupant from a cold temperature (outside) to a more moderate temperature (atria) and then to a warm temperature (occupied space). As a result of this broader tolerance of temperature, the peak loads are reduced and the mechanical equipment runs for a shorter

Inside the sawtooth: a light-filled, airy interior.

Whenever I take people on tours through the building, they say: "Wow, it's really fresh in here, it's like I'm outside, but I'm inside a building". And that's the key to the building: it breathes. The building breathes.

– Mike Godawa, KEEN

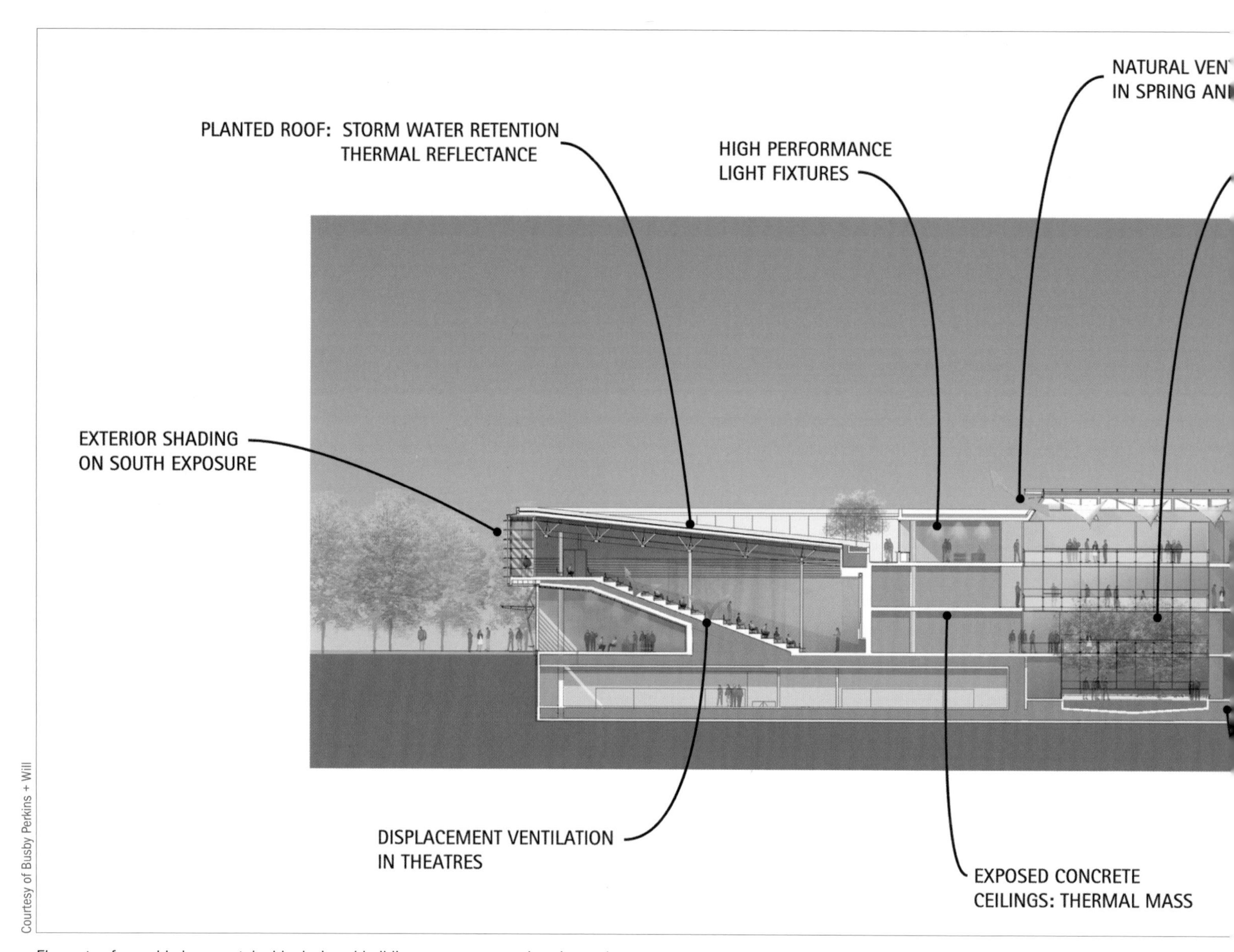

Courtesy of Busby Perkins + Will

Elements of a world-class, sustainably-designed building: warm, open, welcoming and flexible enough to accommodate tomorrow's technology.

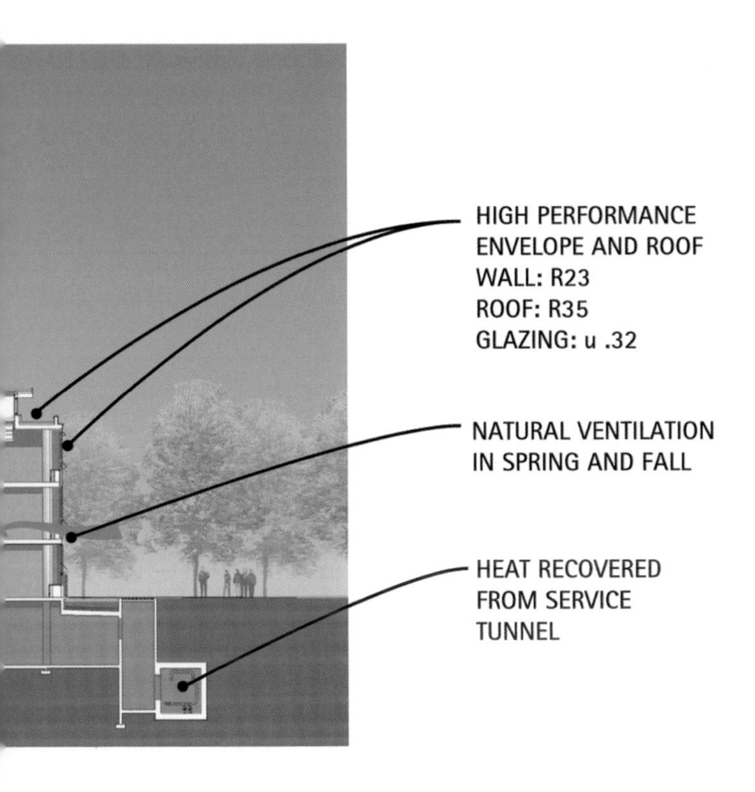

period of time, thus conserving energy.

The University's chilled water and steam plant provides district cooling and heating to the Computer Science Building. A 40°F ΔT (Delta) temperature difference is used for the hot water coils in order to reduce pipe sizes, pump sizes and pumping costs; a 14°F ΔT is used on the chilled water side. A scavenger heat exchanger also recovers waste heat from the condensate to reduce the temperature of the condensate returning to the main plant and reduce the load on the steam to hot water heat exchangers.

Additional design measures were taken to significantly reduce the building's annual energy consumption. These measures include: two large ventilation stacks – solarons – that induce airflow, allowing hot air to escape the building through the tops of the atria; windows on the east and west are located on the sawtoothed walls; south-facing glass is shaded by overhangs; louvres along the principal façade prevent summer sun from driving heat into the central atrium; and walls are well insulated to keep the building cool in summer and warm in winter.

Lighting & Daylighting

Daylighting is central to the success of the building, due largely to windows all around the perimeter and the atria, which allow light to bounce in.

Offices for faculty, graduate students and research staff also contain many sidelights within wooden panels. High-efficiency T8 lighting fixtures also take advantage of the exposed concrete, bouncing light off the exposed white ceilings.

Site

The pre-selected site for the Computer Science Building was relatively tight for new construction, nestled between two existing buildings on the York University campus. As a result, the design team gave careful consideration to where and how to orient various program elements within the building.

Engineering Process

The team's decision to employ passive natural ventilation as a central feature of the Computer Science Building was critical to meeting the University's sustainability goals. Yet, it also required consensus and a willingness between team members and the client to share risk to achieve these green design strategies in an extreme climate. That decision factored into KEEN's knowledge of natural ventilation and hybrid buildings accomplished in the U.K. and Germany and the ability to leverage those technologies and techniques. Still, all the decision-makers knew this venture was something new, and that it was going to be a journey for all involved.

Computer modeling utilized Thermal Analysis Software (TAS), allowing designers to analyze the building from a heat flow perspective and then to predict temperatures for various spaces. This analysis provided valuable information for determining the tolerances and risks in using these techniques.

Another strategic decision, made with the client, addressed transient areas such as the building's main entry points into the atria. Particularly in winter or summer and regardless of outside conditions, these spaces are typically heated or air-conditioned like other spaces in a building. Instead, KEEN recommended that the thermal characteristics of these transient areas be relaxed. York's buy-in to this transient zoning technique has been especially important to energy performance. Other key considerations for natural ventilation included making certain the envelope was superior, and that access and security concerns, arising from windows being left open by building occupants, were addressed.

Finally, at the commissioning stage, KEEN engineers briefed York's maintenance and operations personnel who would ultimately run the building. Their presentation covered all aspects of how the simple, robust system would work, including the components of air handling, mechanical room, underfloor air, fan coils, pumps, heat exchangers, and ventilation stacks. Equally important, the session described the minimalist approach to design as driven by the architecture – and how that design, in effect, would also reduce the required maintenance, eliminate the need to access ceilings, and limit the amount of heating and chilled water piping.

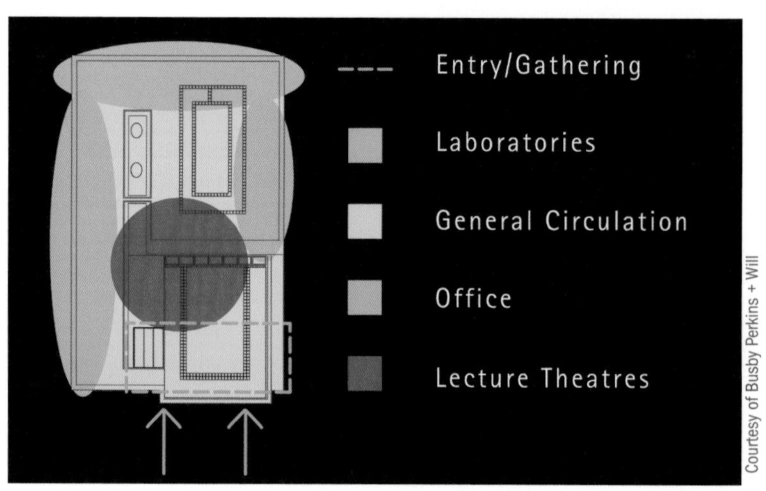

Courtesy of Busby Perkins + Will

KEEN and the project team looked at planned uses within the Computer Science Building to maximize its energy advantages.

There is nothing especially novel in the way York University's Computer Science Building handles green design – no dazzling gizmos or weird new materials – but the three-storey courtyard building is a lesson in putting sensible environmental thinking at the heart of a scheme.

– World Architecture Magazine

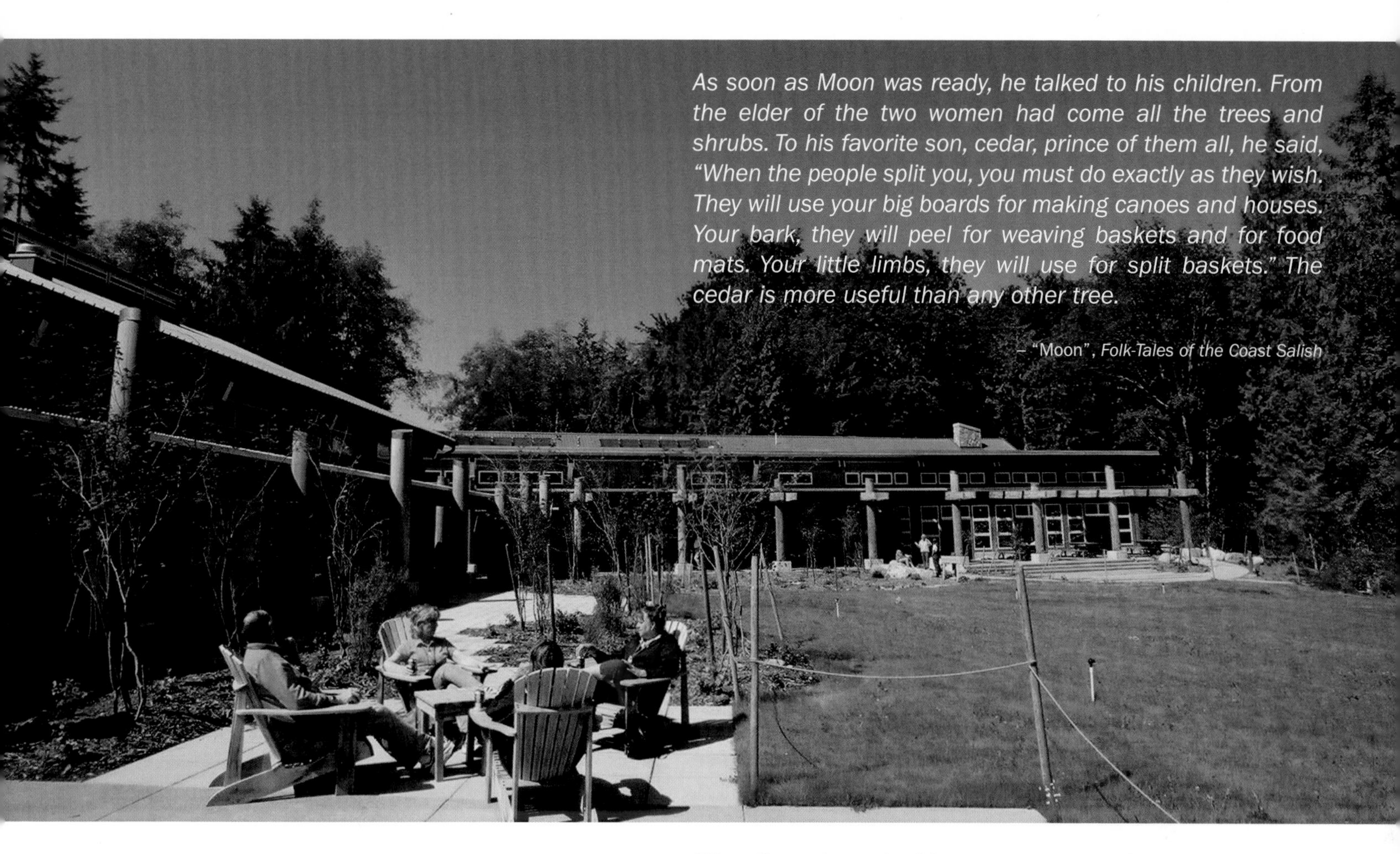

> *As soon as Moon was ready, he talked to his children. From the elder of the two women had come all the trees and shrubs. To his favorite son, cedar, prince of them all, he said, "When the people split you, you must do exactly as they wish. They will use your big boards for making canoes and houses. Your bark, they will peel for weaving baskets and for food mats. Your little limbs, they will use for split baskets." The cedar is more useful than any other tree.*
>
> – "Moon", *Folk-Tales of the Coast Salish*

IslandWood

A School in the Woods

Bainbridge Island, Washington

"I just figured out why this camp was called Camp IslandWood– because it mostly has wood...and it's an island."

Reginald is a Seattle 4th grader and among the nearly 4,000 school children each year who experience the magic of IslandWood, one of North America's most extraordinary outdoor classrooms. Located across Puget Sound from downtown Seattle, this heavily wooded environmental learning center immerses students and their teachers in nature for four days and three nights as part of a unique, hands-on learning experience.

IslandWood's focus is on adventure and discovery and its classroom, a 255-acre property that features a rich variety of ecosystems. During

Copyright © Jim Burns

At IslandWood, off Puget Sound near Seattle, inner-city school children come to learn about Nature – and the power of sustainably-designed buildings.

their time at the center, 4th and 5th grade students, accompanied by naturalists, explore the site – and its forest, ponds and wetlands – on a series of linked trails. All programs integrate science, technology and the arts, helping both children and adults develop a deeper understanding of the relationships between biological and cultural diversity.

At the heart of IslandWood is a mix of campus buildings designed to create a "deep in the woods" outdoor experience for visitors. The sustainably-designed facilities also serve as a model for conservation and community stewardship, where children can operate building controls and electronically monitor their own energy and water use as "conservation apprentices."

"IslandWood was created with the idea that kids from inner city areas didn't often have the opportunity to leave an urban environment and spend time in the natural world," said Debbi Brainerd, the founder of IslandWood. "My idea was to allow them to live in the forest and experience Puget Sound history and native cultures."

Primary Team Members

Owner:
IslandWood – Debbi Brainerd, Founder and Board President

Architect:
Mithun Architects + Designers + Planners, Seattle, WA
Project Team: B. Gregory (Principal-in-Charge), R. Franko, (Project Manager), D. Goldberg (Project Designer), L. Robbins, A. Sturgeon, T. Rooks, C. Kruger, B. Cloward, S. McNabb, J. Harrison, D. Swaab, K. Pirie, C. Dixon, T. Johanson, K. Tam, S. Martin

Engineers:
Mechanical: Keen Engineering, Vancouver, BC – K. Hydes, M. Younger (Engineer of Record)
Structural: Skilling, Ward, Magnusson, Barkshire Inc., Seattle, WA
Electrical: Cross Engineers, Tacoma, WA
Wastewater: 20/20 Engineering, Inc. Bellingham, WA
Civil: Browne Engineering, Inc. Bainbridge Island, WA
Photovoltaics: M. Nelson, Washington State University, Olympia, WA; Schott Applied Power, Olympia, WA
Solar Hot Water: Heliodyne Inc., Richmond, CA

Consultants:
Environmental Consulting: D. Rousseau, Archemy Consulting
Interior Design: E. MacPherson, C. Schmidt, L. Herriot, Mithun
Landscape and Planning: The Berger Partnership Seattle, WA
Master Planning: William Isley, Bainbridge Island, WA

General Contractors:
Educational Core: Rafn Company, Bellevue, WA
Art Studio & Site Structures: Drury Construction, Poulsbo, WA
Staff Housing: Woodside Construction, Kingston, WA
Trails and Site Structures: Sahale, LLC, Seattle, WA

That vision originated in 1997 when Brainerd and her husband Paul, founders of the Brainerd Foundation, learned that a thousand acres of land on the south end of suburban Bainbridge Island was for sale. Familiar with the crescent-shaped swath of second-growth forest, they purchased 255 acres – determined the property would be used for something other than new housing developments.

Washington State had recently mandated that students in the 4th and 5th grades focus on learning about ecosystems, salmon, energy and micro-worlds as well as cultural history. So Brainerd proposed the idea of a not-for-profit, outdoor education program for children, adults and families. Initially called the Puget Sound Environmental Learning Center, IslandWood quickly became a reality through the ideas and involvement of hundreds of children, educators, scientists, artists and community members.

Opened in September 2002, IslandWood's educational core includes the Main Center (Interpretive Center, Great Hall, Administration), Learning Studios, Dining Hall, Art Studio and maintenance building. Visitor accommodations include three lodges and a guest house. North of the educational core is a staff housing area. Shelters, bird blinds, walkways and lookouts are featured throughout the site, all connected by a system of primary and secondary trails.

Environmental Objectives

The Brainerds and IslandWood wanted to minimize the center's environmental impact to the area, which featured a mixed, second-growth forest, uplands, historic cemetery, and the potential for salmon habitat restoration in a stream.

The land around IslandWood and nearby Port Blakely Harbor had been logged extensively starting 140 years ago – and for more than four decades, the Port Blakely Mill Company had the world's largest sawmill.

Initially, scientists and educators spent time on the property to

IslandWood's "deep in the woods" campus of buildings greets thousands of 4th- and 5th-graders and their teachers every year.

Copyright © Roger Williams

The Great Hall's fireplace and chimney stones teach visitors the story of Cascade geology.

The buildings are a textbook. The buildings are things that they can operate. They can learn which direction is north, which direction is south, they can learn how the wind moves, learn how the sun moves, and learn about how to be better stewards of the environment.

– David Goldberg, AIA, Mithun

Project Program and Stats

Location:	Bainbridge Island, Washington
Building Type:	Outdoor Education Center
Size:	70,000 sq. ft. facility
Stories:	One- and two-story buildings
Building Features:	40 buildings and site structures, including the Interpretive Center, Great Hall, offices, Learning Studios, dining/kitchen facility, student and graduate student housing, Arts Studio, maintenance building, and visitor accommodations
Site Issues:	Area was previously logged and was once the site of a village. To minimize site impact, structures were located away from wetland areas and a stream while leaving as many trees as possible and still providing vehicular access.
Completion Date:	2002
Certification:	LEED® Gold Version 2.0 for Main Center, Dining Hall and Learning Studios. Residential buildings constructed in accordance with the "Build a Better Kitsap" environmental construction program.
Cost :	$33 million
Awards:	· AIA/COTE Top Ten Green Projects, 2002 · International Interior Design Association (IIDA), Washington State Chapter, Best of Competition - IN Awards, 2003 · Associated Builders & Contractors of Western Washington Excellence in Construction - Eagle of Excellence Award, Specialty Construction, 2003 · Design Resource Institute International Design Resource Awards (IDRA), Design With Memory, 2002 · American Institute of Architects-Seattle Chapter/IIDA Exhibition, Greenworld – What Makes it Green, 2002 · City of Portland Office of Sustainable Development/American Institute of Architects-Portland Chapter Ten Shades of Green Exhibit, Cascadia Region, 2001

discern what educational "stories" could be shared with children. Biologists were excited by the property's rich variety of ecosystems: sixty-two acres of wetlands, a bog, a four-acre pond, an eagles' roust, emerging forests, and access to a marine estuary park.

Design Overview

Mithun Architects + Designers + Planners of Seattle was selected as project architect, with a team of engineers, planners and consultants that included ten LEED® accredited professionals and various environmental engineering experts.

Planning began with a green visioning charrette and a series of workshops on how to best create an educational center that could be a "magical place for kids." Community members, artists and teachers were also involved at early stages of the design process.

Mithun also suggested that local school children be involved in the design process to help make IslandWood a very child-friendly place. University of Washington landscape architecture students worked with over two hundred-fifty 4th, 5th and 6th grade children to learn what their ideas would be for learning in the natural world. Children's ideas focused on adventure-based learning such as a floating classroom, a suspension bridge, a forest canopy structure, and several tree houses.

Intensive site and resource analysis mapping was conducted to locate campus buildings on previously disturbed sites and to have a minimal impact on sensitive ecosystems. IslandWood's design team examined areas that had not been logged and those with significant vegetation, hydrology, wetlands, existing human use, cultural resources, soils, wildlife, and very steep slopes.

Many sustainable features of the campus are incorporated into lesson plans at IslandWood. Building structures and systems operate as "active" participants in the education process by arousing children's keen instincts for sight, touch, smell, and sound. These structures

The Interpretative Center is also geared for learning: daylight, natural ventilation and recycled building materials were combined in the space to arouse students' sense of sight, touch, smell and sound.

and systems serve as teaching tools, designed to tell the stories behind the site's green features such as passive solar design and use of natural daylight, as well as natural ventilation, photovoltaic (PV) panels, solar hot water, radiant heating, recycled materials, rainwater collection, constructed wetlands, and native and edible plants across the campus.

The columns of the Great Hall were carved by Coast Salish Native American craftsmen. Numerous local craftspeople and artists were also commissioned to produce environmentally-sensitive furniture and artwork for the campus. Careful thought was given to fireplaces and chimney stones, rainwater cisterns, and artist-made building materials to facilitate an understanding of ecological connections and interactive hands-on learning – for example, using local stones in the construction to explain Cascade geology.

Copyright © Roger Williams

A Living Machine™ and greenhouse is yet another interactive classroom at IslandWood – designed to treat and recycle all greywater and blackwater on site, realizing a 70 percent to 80 percent potable water savings for the facility.

Copyright © Jim Burns

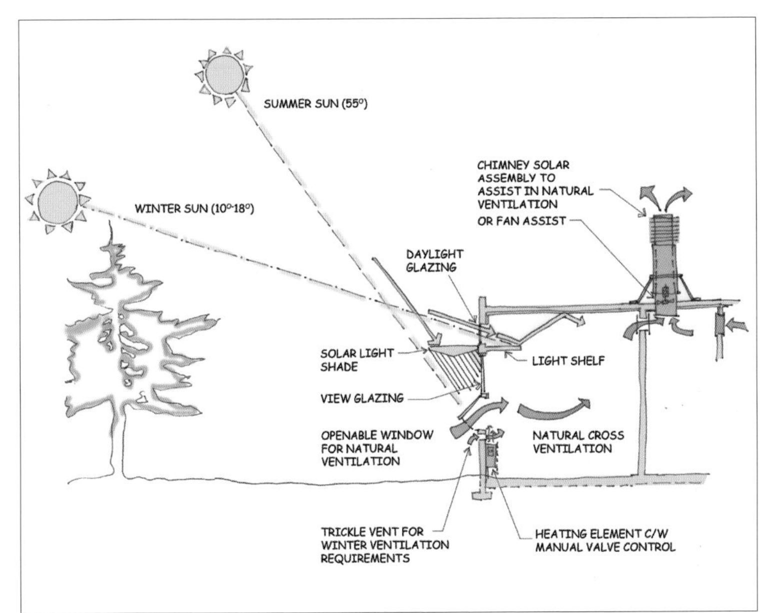

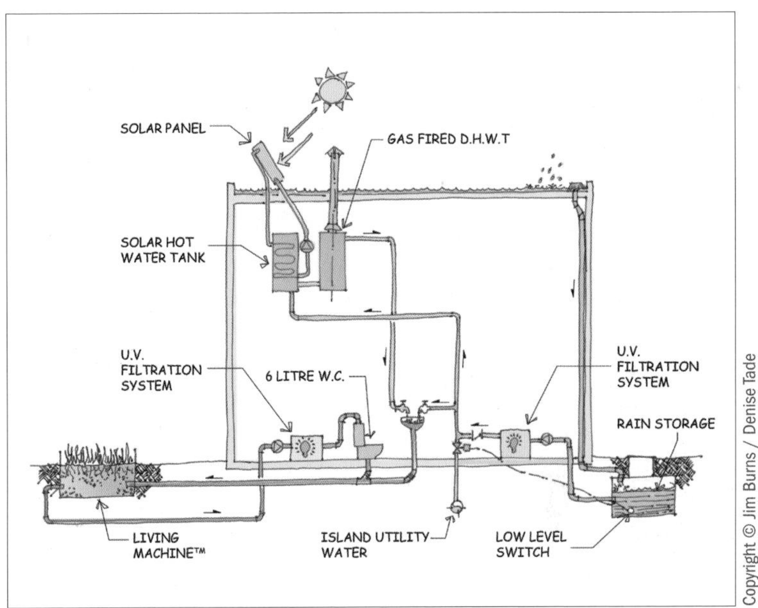

KEEN's original concepts for air and light (left) and water (right) schemes remained consistent with the Foundation's limited operating budget, environmental objectives and educational mandate.

Solar meadows, created within the dense forest, allow direct sunlight into the buildings and provide a program-required play area for children, while the felled trees were milled for siding and trim. Open air site structures also provide a dry place in the damp Northwest forest for writing, drawing, and outdoor field experiments. A Living Machine™, located inside a greenhouse at IslandWood, serves as a natural wastewater treatment system and interactive aquatic science classroom.

All buildings at IslandWood are designed for long life and low maintenance in keeping with the center's limited, non-profit operating budget. Steel and wood roof structures are left exposed and unfinished, as are structural slabs-on-grade in most areas. Most interior materials were left unfinished and in their "natural" state. Where paints, coatings, adhesives, or sealants were required by code, only durable, low VOC finishes were used. Walls are insulated with 100 percent recycled content cellulose and 57 percent of wood in the project is FSC certified.

Other sustainable materials include: 50 percent fly-ash concrete, 100 percent recycled cellulose insulation, 95 percent recycled content carpeting, recycled glass bathroom tiles, recycled rubber flooring, strawboard, formaldehyde-free MDF, cork, bamboo, salvaged fir flooring and recycled plastic toilet partitions. The flooring and interior trim, and much of the siding and custom furniture, was assembled from site-harvested cedar, fir, alder, maple and hemlock.

The Welcome Center and Great Hall (16,000 square feet) include timber trusses salvaged from a mine in Montana – including a ninety-two foot, 4-1/2-ton Douglas fir beam, milled at the Port Blakely Sawmill around 1900 and later donated to IslandWood to serve as an element in the primary roof truss.

In addition, each classroom in the Learning Studios (7,500 square feet) has a different sustainable flooring material – cork, bamboo, recycled rubber, and concrete – and a different sustainable countertop

Most of the steel, wood and concrete materials on buildings, such as the Interpretive Center (above), were left unfinished and in their "natural" state.

surface, such as recycled yogurt container composite material, soybean and sunflower seed bio-composite material, or high fly-ash-content concrete. In the 7,500-square-foot Dining Hall, roof rainwater is collected in cisterns for use in landscape irrigation. Bathrooms feature recycled glass tiles in flooring and walls and waterless urinals in the men's restroom save a gallon per flush. Art Studio walls are constructed from straw bales.

Building Performance

As an important educational tool at IslandWood, students can monitor their water and energy usage in various buildings through an integrated computer data network. Over the course of their four-day stay at IslandWood, children can quantify and track their improvement and learn conservation strategies.

Energy monitoring and measuring systems, including CO_2 monitoring, are interconnected to a central database using the International Performance and Verification Protocol (IPMVP).

MEP Design Process

In meeting the rigorous LEED® Gold criteria, KEEN's challenge was to provide systems that would set a precedent in natural ventilation while maintaining documented health and comfort levels for staff and students. The area's mild climate, strong cooling breezes and plentiful natural shading offered unique conditions to benefit from savings in energy and capital costs. In designing the mechanical systems, KEEN followed the LEED® Comfort Parameters that allow for a wider range of temperatures than traditional design expectations. The design team incorporated a variety of solar strategies into the design to minimize solar heat gain in summer and maximize it in winter – such as southern building orientation, strategic overhangs, shade devices, and multiple sources of daylight with clerestories and operable skylights.

In the Main Center, all cooling, solar heating and natural ventilation is provided by a combination of operable windows, photovoltaic roof fans

I had a great time…I liked the cabins a lot because it had bunkbeds and a couch. The cabins were warm and the place felt like home. In my study group, we learned about plants and we did art projects. I will always remember the fun time I had here.

– Carmen, Kitsap County 5th grader

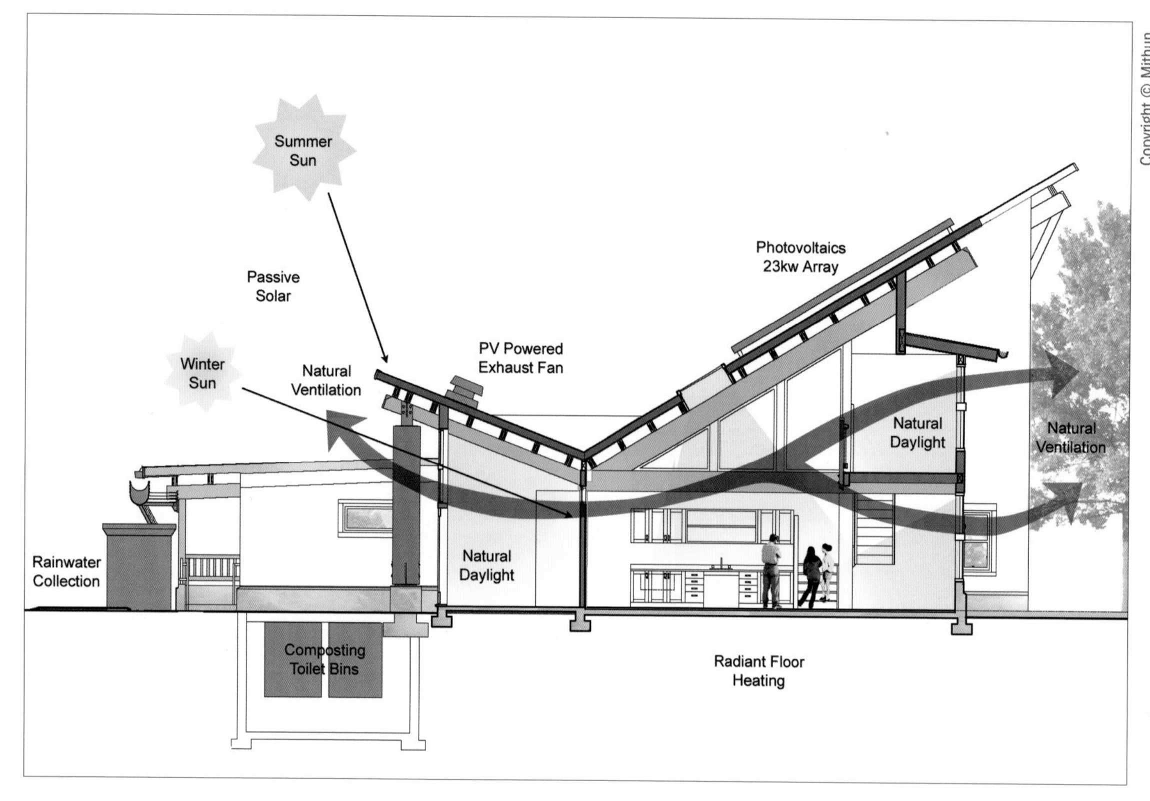

Light, air, water: Mithun's rendering of the Learning Studios – at the heart of IslandWood – highlighted the many possibilities of sustainable design.

and mechanically-operated ventilating louvers. Retractable shading canopies are used to reduce the summer cooling load. The Great Hall, sized for several hundred people, features operable windows on the perimeter and skylights in the middle, plus operable louvres.

The Learning Studios optimize solar gain as a result of solar meadows to the south. The building is bio-climatically responsive by passively receiving solar heat and daylight, and by optimizing solar harvesting on the south-facing butterfly roof face.

Indoor Environmental Quality

Two-week building "flushouts" were performed at IslandWood prior to occupancy. In maintaining indoor air quality, skylights and operable windows replace air conditioning. A skylight with mechanically-operated louvers provides solar heating in winter and passive ventilation in summer. Photovoltaic powered fans assist with room ventilation.

Water and Waste Design

Civil and mechanical engineering firms, as well as architectural firms, worked together on a strategy to process all building wastewater on site to tertiary standards, requiring no connection to a municipal sewage system. Wastewater from the Sleeping Lodges is naturally treated with a subsurface flow constructed wetland. Water from the three main buildings drains to a Living Machine™, an on-site tertiary wastewater treatment facility.

IslandWood's Living Machine™ process is an important part of the educational program, as it teaches students through direct observation about nature's own filtration capacity. Located inside a greenhouse, the system treats an average daily flow of 3,000 gallons through oxygenation, aquatic plants, snails, fish, and anaerobic and aerobic bacteria. All the waste is fully recycled and reclaimed water is pumped back to the main facility for subsurface irrigation and toilet flushing at the Dining Hall and Welcome Center bathrooms.

Additionally, all buildings include low-flow water fixtures. Men's room no-flush urinals at the Dining Hall and composting toilets at the Main Center and Learning Studios eliminate water use, saving 1 to 2 gallons of water per flush. Exterior concrete cisterns collect roof rainwater for boot washing and landscape irrigation. The Visitor Lodges are plumbed for greywater re-use.

Heating and Cooling

The project team's initial energy goal was to meet the LEED® 1.0 energy credit requiring no active heating or cooling for eight months of the year.

A "butterfly" roof configuration optimizes passive solar gain and still provides a large surface for solar hot water or PV panel installations at the Dining Hall and Learning Studios. R-40 roofs and R-19 cellulose insulation walls combine with high efficiency windows to achieve an efficient building envelope. Buildings also feature high efficiency in-slab radiant heating systems and high-efficiency hydronic boilers. Operable windows, PV and electric powered ceiling fans assist with cooling.

Lighting & Daylighting

Daylighting is an important element of IslandWood, achieving views for 90 percent of all spaces. Skylights and tall north-facing windows in the Welcome Center and Great Hall eliminate the need for electrical lighting during the day. In the administrative offices, daylighting is controlled with natural wood blinds, retractable canopies and

IslandWood's Learning Studios (classrooms), located adjacent to solar meadows on the south, take full advantage of daylight.

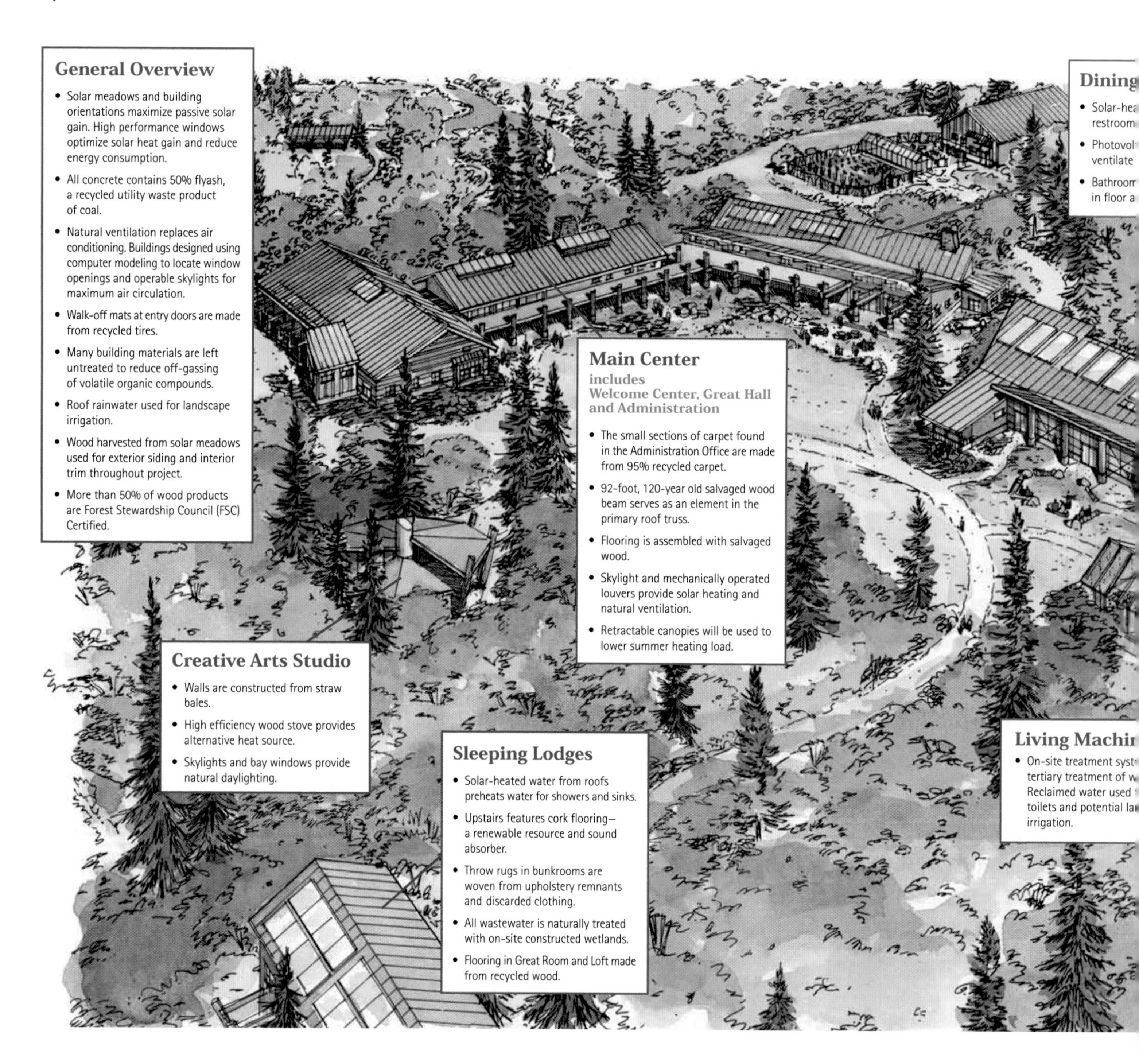
General Overview
• Solar meadows and building orientations maximize passive solar gain. High performance windows optimize solar heat gain and reduce energy consumption.
• All concrete contains 50% flyash, a recycled utility waste product of coal.
• Natural ventilation replaces air conditioning. Buildings designed using computer modeling to locate window openings and operable skylights for maximum air circulation.
• Walk-off mats at entry doors are made from recycled tires.
• Many building materials are left untreated to reduce off-gassing of volatile organic compounds.
• Roof rainwater used for landscape irrigation.
• Wood harvested from solar meadows used for exterior siding and interior trim throughout project.
• More than 50% of wood products are Forest Stewardship Council (FSC) Certified.
Main Center
includes Welcome Center, Great Hall and Administration
• The small sections of carpet found in the Administration Office are made from 95% recycled carpet.
• 92-foot, 120-year old salvaged wood beam serves as an element in the primary roof truss.
• Flooring is assembled with salvaged wood.
• Skylight and mechanically operated louvers provide solar heating and natural ventilation.
• Retractable canopies will be used to lower summer heating load.
Creative Arts Studio
• Walls are constructed from straw bales.
• High efficiency wood stove provides alternative heat source.
• Skylights and bay windows provide natural daylighting.
Sleeping Lodges
• Solar-heated water from roofs preheats water for showers and sinks.
• Upstairs features cork flooring—a renewable resource and sound absorber.
• Throw rugs in bunkrooms are woven from upholstery remnants and discarded clothing.
• All wastewater is naturally treated with on-site constructed wetlands.
• Flooring in Great Room and Loft made from recycled wood.
Dining
restroom
ventilate
Living Machi
tertiary treatment of
Reclaimed water used
toilets and potential
irrigation.

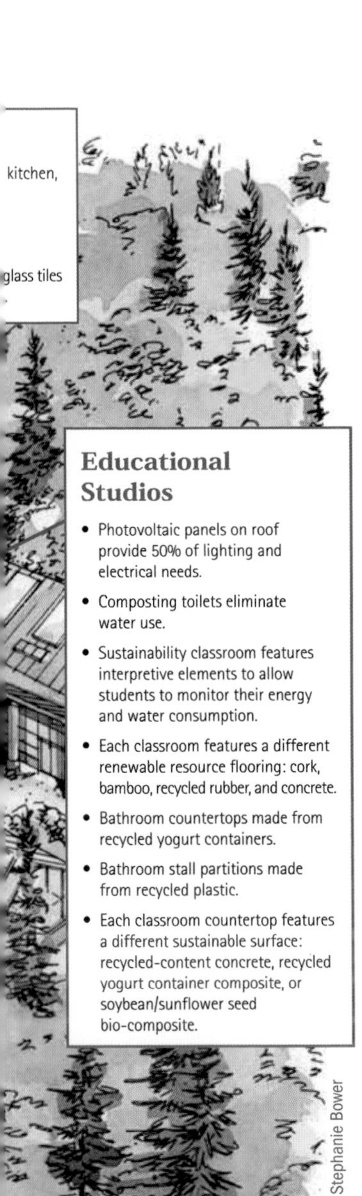

Educational Studios

- Photovoltaic panels on roof provide 50% of lighting and electrical needs.
- Composting toilets eliminate water use.
- Sustainability classroom features interpretive elements to allow students to monitor their energy and water consumption.
- Each classroom features a different renewable resource flooring: cork, bamboo, recycled rubber, and concrete.
- Bathroom countertops made from recycled yogurt containers.
- Bathroom stall partitions made from recycled plastic.
- Each classroom countertop features a different sustainable surface: recycled-content concrete, recycled yogurt container composite, or soybean/sunflower seed bio-composite.

mechanized skylight blinds. T-8 lamps, T-5 fluorescent lighting and dimming ballasts with photoelectric controls in all occupied rooms further reduce lighting energy demands.

Renewable Energy

On top of the Learning Studios, a 23kw photovoltaic installation powers over 50 percent of lighting and electrical loads. At the time of completion, this PV array was the largest installed in Washington State. Attached to the PV panels are a series of "attic fans," with dampers underneath, that can kick on automatically to vent excess heat in the classrooms. A photovoltaic exhaust system is also located in utility spaces. Rooftop solar panels on the Dining Hall provide domestic hot water for the kitchen and showers in three Visitor Lodges, delivering 50 percent of annual demand.

Site

The design team used eco-mapping to determine the appropriate building sites for IslandWood's educational core. Their survey catalogued every tree over eight inches in diameter across a sixteen-acre forested area. This survey allowed the team to locate individual structures around significant groupings and mature individual specimens – as a result, individual building sites take up only 6 acres out of 255. A concentrated analysis further examined optimum building orientation, shape and configuration.

Engineering Process

Mithun, KEEN and the entire design team recognized the importance of early and constant collaboration in creating IslandWood, as well as regular contact with the client. That close collaboration – in effect, a joint discovery process – yielded early solutions for building shape and orientation, natural cooling, ventilation strategies and air movement. This iterative process resulted in the "butterfly" roof form for several buildings, to optimize passive solar gain and enhance natural ventilation.

Far Left: Winner of a 2002 AIA Top Ten Green Projects Award, IslandWood was also the first project in the state of Washington to receive a Gold LEED® rating - and the third Gold in the United States and fourth in the world.

Bert Gregory, now president of Mithun, also recalled early discussions with KEEN to determine the best time-of-use building strategies. Those buildings on campus predominately used during the day, such as the Learning Studios, were sited and designed for direct solar radiation. For night-time use buildings such as the lodges, the team worked through the calculations and costs to develop the best way to store heat from the day and release it overnight. "At the end of day," said Gregory, "it was better to take an extremely compact strategy as opposed to a big, long-building, south-facing strategy. I think the 21st Century is really all about this integrated design approach, where you have your key engineers on board at the project conception to make sure that as you move forward, the building is formed incorporating all these different issues – and you innovate from that."

Buildings at IslandWood were modeled using TAS (Thermal Analysis Software) software to optimize natural ventilation. This tool was essential in developing the building form and fenestration so that thermal comfort could be achieved without air conditioning and to minimize heating. Iterative 3D models and revisions to the architectural design – including window sizes, orientation, window shading and shading co-efficients – continued throughout schematic design.

Because IslandWood is a not-for-profit venture and privately funded, difficult economic choices had to be made during the entire design process to balance issues of sustainability with programming and aesthetics. Exposed concrete slabs replaced wood floors. Tiles were removed from many bathrooms. FSC-certified wood siding was cost prohibitive. And more sophisticated rainwater harvesting systems were put on hold for future installation.

> When we began the project, one of the ideas was to use architecture as education, so that the buildings did more than contain programs or hold people – they were an opportunity to help teach about sustainability, about the natural world.
>
> – Debbi Brainerd
> Founder and Chairperson of IslandWood

Still, most of the project's sustainable goals were achieved by working closely as a team and trading off costs. For example, because buildings did not rely on fully-ducted mechanical systems, the client could afford better insulated, operable windows. In the end, the team delivered the most complex of the buildings for less than two-hundred dollars per square foot while at the same time being 25 percent more energy efficient than typical construction in the area.

Welcome to IslandWood: It took a true integrated design process and project team to make this 255-acre "School in the Woods" a reality.

Library Square
Canada's Colosseum

Vancouver, British Columbia

When Vancouver's newest civic symbol officially opened on May 24, 1995, *The Vancouver Sun* summed up nearly everyone's views of Library Square: "Love it, hate it: conventional wisdom is that you're looking at the heart of our 21st-century downtown."

The newspaper then declared that the Vancouver Public Library's new central branch would join the list of the city's most impressive landmarks. The multi-building complex occupies a complete city block and, at nearly $160 million, was the largest capital project ever undertaken by the City of Vancouver. In addition to the library, it contains provincial and federal government offices, a variety of shops and cafes, a rooftop garden, and below-grade parking.

Primary Team Members

Owner:
City of Vancouver

Architects:
Moshe Safdie & Associates, Boston, MA
Project Team: M. Safdie (Principal-in-Charge), P. Matthews, M. McKee

Associate Architect: Downs/Archambault & Partners, Vancouver, BC
Project Team: D. Galpin, R. Beaton

Engineers:
Mechanical: Keen Engineering Co. Ltd., Vancouver, BC - B. McCarry (Engineer of Record), B. Johnston (Project Manager)
Structural: Read Jones Christoffersen Ltd., Vancouver, BC
Electrical: Schenke/Bawol Engineering Ltd., Vancouver, BC

Consultants:
Construction Management: NW Fletcher & Associates, Vancouver, BC
Landscape Architecture: Cornelia Hahn Oberlander, Vancouver, BC

General Contractor:
PCL Contractors Pacific Inc., Vancouver, BC

Simplicity was a hallmark of the design. A nine-story rectangular structure, lined in glass along its east side, contains the library's book stacks and services. This central core is a straightforward grid of continuous floor plates and structural columns of exposed concrete. Reading rooms with vast windows are located at the periphery of the library in the concrete outer shell, connected by a series of narrow bridges that cross a five-story atrium and the concourse below.

The library's curved "ghost wall" recalled – for many Vancouverites – Rome's famous Colosseum. A second wall rises up at the south end of the site and sweeps toward a twenty-one-story office building at the northeast corner. The office building, leased for government offices, is clad in precast concrete panels two stories high with the same granite aggregate as those of the library.

At the southwest corner, the precast walls open out to form a street-level piazza and entrance to the soaring, glass-roofed concourse that connects Georgia and Robson streets. Known as the Promenade, the atrium-area has become an indoor urban square – where people sit and read in the midst of shops and cafés facing the multi-level library.

The origins of Library Square date back to November 1990, when a city referendum asked residents whether a new central library was needed. A two-stage, international design competition followed, with the short-listed architects asked to produce an "Expression of Vision" for the new library project.

In August 1992, Vancouver's City Council announced the winning submission from Moshe Safdie & Associates of Boston with Vancouver's Downs/Archambault Partners. Citizens had been polled on their responses to the three different designs in the competition. Seventy percent of respondents favored the organization, circulation and lighting of Safdie's design – as did the competition jury, made up of civic officials and architects.

Library Square also quickly became Vancouver's most controversial new building. Heavy press coverage referred to it as "a radically ancient design" (*British Columbia Report*, April 27, 1992), "Library designs have Vancouver abuzz" (*Toronto Star*, March 14, 1992) and "Brave new world" (*Vancouver Sun*, Dec. 10, 1994). Many architects and critics dismissed it as a "grandiose folly", resembling a small-scale Roman Colosseum and an "overblown cardboard stage set."

Yet Safdie said it was not Imperial Rome he had in mind – but a way of adapting the traditional reading room to today's technology. The October 1995 issue of *Architecture* magazine reported his belief that the working library of a big city in the 1990s "should be 'inviting and transparent' and incorporate the latest in computer technology", and that he wanted "to evoke traditional values of learning and citizenship on the outside of the building, to shape 'a civic place with a sense of civic identity.'"

Controversial or not, most Vancouver residents fell in love with their brash, elliptical-shaped library. The new library replaced an existing 130,000-square-foot space with a building three times that size. The improved facilities featured 1,200 reading galleries and arcades circling the seven floors, along with 250 computers, a literacy center and language lab. Nearly 1.5 million items are housed in the central library's collection: in addition to books, periodicals, microforms and recordings, more than 200,000 historical photographs are inlcuded.

As a member of the winning project team, KEEN's contributions – an underfloor air design system, thermal mass, natural ventilation and an innovative ice storage scheme – were also integral to the success of Library Square, resulting in ASHRAE's 1998 International Technology Award for Institutional Buildings.

Library Square also marked a new and significant shift in focus for Vancouver's Downtown South, which had not historically been a pedestrian district. The introduction of a major public building to the area helped to extend the vitality of the existing downtown core, spawning new condos, hotels and the Ford Theatre for the Performing Arts – also designed by Safdie.

Moshe Safdie's "Expression of Vision" for Library Square relied heavily on light, mass, history and technology to integrate a major office and civic structure into a vibrant new public space for Vancouver.

"In the five months it has been open," stated *Architecture*, "Library Square has already proved to be a considerable success. To start with, Safdie has adopted an intriguing site plan in a district of Vancouver

Copyright © Jim Burns

Five-story light wells, spanned by bridges over the Promenade, separate the library's bookstacks in the rectangular main building from reading and study rooms located in the curving, freestanding outer shell.

It was no longer a question of keeping track of all books issued from the press: that was already recognized to be impossible. The difficulty became one of keeping track of lists of books.

– from *A World Bibliography of Bibliographies*, the first book to arrive at Vancouver's new public library at Library Square

Project Program and Stats

Location:	Vancouver, British Columbia
Building Type:	Mixed-Use Building Complex
Size:	398,000 sq. ft. library; 290,000 sq. ft. office space; 30,000 sq. ft. retail area
Stories:	9-story library; 21-story office tower, 5-story atrium
Building Features:	Vancouver's central public library, government office tower, retail and service facilities, rooftop garden, 700 below-grade parking spaces
Site Issues:	Introduced a major civic building into an area that had not historically been a pedestrian precinct
Completion Date:	May 1995
Cost:	$156.8 million (CAN)
Awards:	· ASHRAE Technology Award - First Place, 1998 · BC Hydro Power Smart Award of Excellence, 1996

that badly needed visual brightening and enhanced community facilities....Visible from several approaches to downtown, the new building had to make a big, sculptural statement."

Environmental Objectives

The project's stated environmental objectives focused primarily on its green roof. The rooftop garden for Library Square was designed to imitate British Columbia's Lower Mainland. The 33,000-square-foot project was one of the first of its kind. Created by Cornelia Hahn Oberlander, the green roof features ornamental blue grasses and deep green kinnickinnick in a pattern that replicates the flow of the Fraser River. The garden also serves to increase biomass and bird habitat in the city, reduce airborne pollutants, store and delay stormwater runoff, and reduce the building's heating and cooling requirements.

The Library Square complex has revitalized Vancouver's Downtown South district with a new office tower (left), retail and service facilities, below-grade parking and the elliptical-shaped Central Branch of the Public Library (right), featuring a 33,000-square-foot green roof and public garden.

Design Overview

As stated in the competition brief, the architect's principal design challenge was to integrate a major office building with a major civic structure, while retaining a generally "inviting and transparent" atmosphere. Safdie accomplished this goal through a layering process. At the heart of Library Square is the library, a rectangular glass structure cradled inside a curving precast concrete wall. Safdie differentiated the bookstacks from reading areas by placing the books in the central building and locating reading and study rooms in the freestanding outer shell to take advantage of natural light and outside views.

"I looked at the great reading room of the New York Public library and decided it was obsolete," he said in an interview with The Vancouver Sun.[1] "I could see dozens of computers on the tables and decided that people today need more privacy and sense of seclusion while they work and yet they should still be a part of the community."

Safdie's solution was the transparent "living study wall" that surrounds the main core of the library. Patrons can select a book and walk across a bridge to a reading area that looks back into the library and out over the cityscape. Five-story light wells highlight the separation between the library wall and the curving outer wall. "It bridges a kind of high-tech, futuristic experience at the same time as it provides the comfort and familiarity of something that is both ancient and traditional," he remarked.[1]

Below Library Square's Promenade – an enclosed, unheated concourse beneath a glass roof – is where the shops and cafés are located and it provides a walkway between two streets. Elliptical in plan and four stories high, the precast concrete wall was constructed with a granite aggregate of terra-cotta hue and touches the beige concrete envelope at many points.

In an interview with the CBC at Library Square's grand opening, Safdie noted: "I would like to think that this becomes a great gathering place in the city, I mean a real address. This becomes the cultural sector of downtown, this becomes, like in the old tradition, a great meeting place in the city. We don't have these spaces anymore."[2]

Stepping up from the eastern edge of the complex is the office tower, which Safdie incorporated with the other structures by varying its fenestration to adjust to the library's curved wrapper. Lastly, the roof of the library serves as a large public garden, viewable only from adjacent towers.

MEP Design Process

With Blair McCarry as lead designer, KEEN's primary objective was to find engineering solutions that fully supported the architecture – systems that were "subtly functional" and would not become an obtrusive part of the building. They examined a number of alternatives and settled on design features that blend into the architectural scheme while complying with the requirements of ASHRAE/IESNA Standard 90.1. To retain the clean lines of the library's arched ceilings (the exposed underside of the structural concrete slab), cable trays, electrical wiring and the air supply were all located within the carpeted access flooring. Effectively, the sprinkler heads for each level were placed in the access floor on the next level above. Among the bookstacks, simple strips of lighting reflect against the architectural ceilings.

The underfloor system supplies air to the library's interior areas while four-pipe fan coil units serve the perimeter. KEEN recommended installation of an access floor – not commonly used in North America at the time – giving the library the flexibility to accommodate future changes in computer technology with a minimum of internal reconfiguration costs. The floor plenum, between the access floor panels and concrete floor, is used as the supply air plenum. The access flooring is two feet deep; each removable 2-foot by 2-foot floor tile gives technicians ready access to computer cables and

communications wiring beneath the paneled walking surface. The cost for the underfloor system was financed by eliminating the traditional architectural ceiling.

Library Square's mechanical engineering solutions reflect several other innovative engineering approaches as well. A low-temperature, ice storage system creates a stable, drier environment in the library's Special Collections area by lowering the relative humidity of air for improved preservation of rare materials – which include books, photographs and other fragile materials. The ice storage system also resulted in lower capital costs (for installation) and operating costs.

Library Square's Promenade and its soaring glass atrium has become a favorite gathering place for thousands of Vancouver residents. Filled with shops and places to read or eat, the enclosed, unheated concourse was designed to maximize natural daylight and ventilation.

Design of the building envelope included pairs of columns in the arcade, which provide substantial shading to the interior of the building – particularly for the two lower floors where glazing is recessed ten feet. The exposed concrete ceiling and floor takes advantage of dynamic thermal storage – also not common in North America's modern air conditioned buildings. Temperature sensors were cast into the slab to ensure that correct temperatures are maintained.

I felt it should be a building that has a sense of mass about it on one hand, that at night it becomes totally transparent and you just see right through the building. It becomes almost weightless.

– Moshe Safdie, Moshe Safdie and Associates [3]

The building also features energy-efficient supply and return fans that utilize aerodynamic inlet and outlet silencers that produce significant static regain from the high annular velocity of the axial fans. This

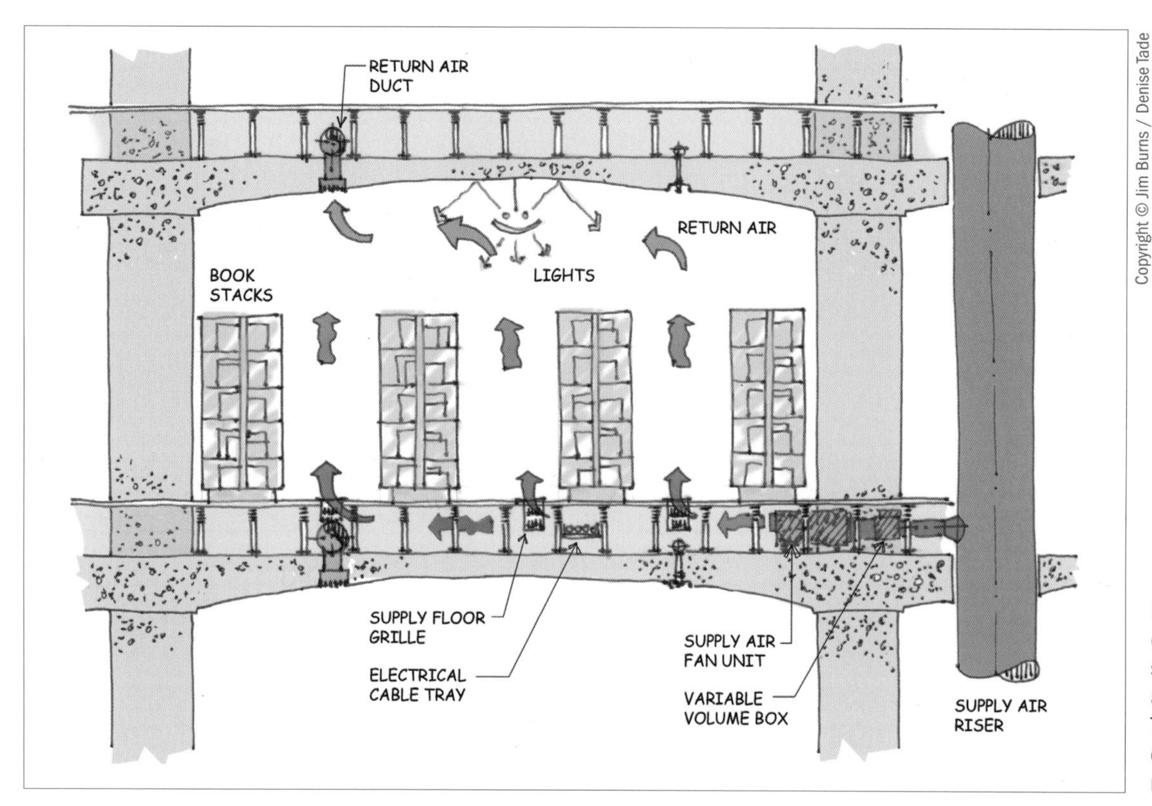

KEEN's concept sketch identified underfloor displacement ventilation as the best solution for supplying fresh, low-pressure air throughout the library's occupied areas. This scheme also significantly reduces cooling loads, allowing warm air to rise naturally off people and computers.

system essentially recovers one inch of static pressure, about 25 percent of the fan pressure, and reduces fan power requirements throughout the project.

Indoor Environmental Quality

The indoor air quality of the mechanical system operating within Library Square was a major consideration for the project. The underfloor air supply acts as a displacement ventilation air system where low-pressure air gently rises from outlet grilles in the floor to return grilles in the ceiling to provide clean, fresh air in occupied areas. Therefore, people-generated pollutants tend to rise and are not mixed in the room.

The underfloor air distribution system operates as follows:

Low temperature air (7°C/45°F) from the main air-handling unit is ducted to the floor plenum to VAV boxes and floor fan units; these units blend filtered local return air and low temperature, VAV-controlled air that discharge into the open floor plenum at approximately 17°C/63°F. Very high induction swirl floor diffusers (8-inch diameter) create a comfortable upward motion of air which supports the natural motion of heat rising from people, computers and lights. Warm stratified return air is then collected through openings in the structural slab into return ducts in the floor plenum of the floor above.

The promenade/concourse is not heated or cooled directly. Natural ventilation is achieved with operable windows and roof openings in summer. In the winter, relief air from the library is transferred into the concourse to temper it and allow the library to meet Vancouver's energy by-law, one of the few municipalities in Canada with an energy by-law.

Superior outdoor ventilation rates are maintained by continuously monitoring the supply volume of the outdoor air and the CO_2 levels in the return air. There is no exposed fiberglass insulation in the supply ducts. Fiberglass-free pre-filters and bag filters (85 percent efficient) are used. Space is provided for additional absorptive air filters as an added environmental benefit to building users if the outside air registers unacceptable levels of air pollution.

In researching archival book requirements for the library's Special Collections, McCarry determined that a dryer environment would lead to longer book life. KEEN reviewed a low temperature air supply system (42°F) with ice storage, designed to deliver relative humidity in the low 40 percent range. The low temperature ducts are significantly smaller than conventional ducts, resulting in $250,000 cost savings and $20,000 lower annual operating costs. Furthermore, the ice storage system fit in a conventionally-sized mechanical room. This low temperature air and ice storage technique had not been used previously in the Vancouver area.

Water and Waste Design

Low-flow fixtures are used throughout the building complex.

The transparent "living study wall" (left) surrounds the main core of the Public Library, with views northeast to the twenty-one story government office building.

Heating and Cooling

The underfloor air system design also significantly reduced cooling loads through the use of stratification. KEEN engineers determined that heat from lights and computers would not greatly affect the room cooling load, compared to its effect in conventional buildings. This approach conditions the occupied zone below six and one-half feet and allows heat to accumulate at the upper stratified zone. Displacement ventilation also supports the natural rising of warm air off the equipment and occupants. As this heat rises, it does not become a space cooling load and it ultimately assists in reducing the amount of energy spent on air conditioning.

The library's exposed concrete ceiling and structural slab offered an excellent application for thermal storage. This storage was accomplished with the heat accumulated in the building structure during the day. The heat is then purged overnight using fans

Copyright © Jim Burns

that operate on 100 percent cool outdoor air. To ensure correct temperatures are maintained, temperature sensors were cast into the slab. Thermal storage further reduced the peak cooling load during the day and reduced the amount of energy required to cool the building.

The ice storage system represented another creative design approach by KEEN, delivering low-temperature air for drier interior conditions while reducing peak electrical demand. Instead of using a large chiller for the cooling system, the Library Square system features a small chiller and an ice-making plant – creating cool air through the night that is released during the day. Supply and return air ducts for the library were trenched under the parking levels.

Lighting & Daylighting

KEEN provided input to the lighting scheme for Library Square. The building's entry concourse was designed with a single-glazing wall for the library and significant exterior glazing and structural skylights over the atrium. There are solar shading devices along the perimeter of the building.

Within the library's central pavilion, indirect lighting fixtures suspended from the wide, shallow ceiling coves provide comfortable, high-quality lighting down to the bottom row of the bookstacks. A mock-up and testing were conducted to verify that these areas had excellent lighting quality.

Lightwells and skylights glazed with solar absorbing glass were strategically placed to maximize the amount of natural light for reading areas. Lightwells at the corners of the library also serve as glass return air plenums for the mechanical system.

Site

Shadows were a concern for Library Square's original, planned location, as the twenty-one story tower would place the library structure in shade. To correct this shading problem, the tower was shifted diagonally across the site. In addition, one level below grade was eliminated to save on excavation costs.

Engineering Process

At the project's inception, KEEN engineers participated in a series of working sessions with architects from Moshe Safdie & Associates and Downs/Archambault & Partners. Safdie's primary design criteria were: free open space to allow effective library usage; integration with the rigid frame concrete seismic design and major perimeter beams; wiring flexibility to meet future technology changes; and high ceilings to accommodate the large floor plates.

Following a thorough analysis of options, the project team arrived at a mechanical system consisting of the underfloor air distribution system, the ice storage and low temperature air system, and dynamic thermal storage. In particular, team members agreed the underfloor air scheme was the best way to highlight architectural elements like the arched ceilings by eliminating fixtures and mechanical components. In addition, the higher cost of the raised floor was offset by savings in the mechanical and electrical systems.

KEEN performed basic, spreadsheet-based calculations to size the underfloor air distribution system. Computer simulations on the ice storage system estimated operating cost savings of $20,000 per year from reduced fan power and electrical demand charges. In addition, engineers completed an airflow study to verify the location of outdoor air intakes.

Excavation began at the Library Square site in January 1993. Given a fixed capital budget of $100 million and the city's mandate that the library would open on schedule with public festivities in May 1995, the project team relied on a twenty-six-month, fast-track approach to construction. This schedule included four major components: site preparation and excavation; the underground parkade; the office tower; and the library. The library was under design while significant components were under construction.

Our goal was to find engineering solutions, in terms of occupant comfort and air quality, that fit the building. It was also important that the designs were nominally visible, that they blend in, to respect the powerful architecture of Library Square.

– Blair McCarry, KEEN

The unique architecture and mechanical design of Library Square presented several interesting challenges for KEEN engineers. The large wall of structural glass separating the library from the concourse has single glazing, whereas Vancouver's energy code called for double glazing. As a solution, McCarry viewed the glass wall design as an "exterior buffer zone" – as a single-glazing subclimate that would still meet ASHRAE standards. Even with no direct heating or cooling, the area was still able to maintain a temperature of 12°C/55°F and accommodate the code requirements.

Another, unexpected, lesson for KEEN occurred in Library Square's second year of operation when a leak appeared in the tank of the ice storage system. In their initial designs, engineers had found creative ways to eliminate approximately one-third of the conventional cooling load for the building. As designed, the system's supply air was then divided equally between the chiller and ice tank running simultaneously. Suddenly, however, the chiller became the sole source of cool air for the entire complex when the ice tank failed.

"What we learned, though, was that the building actually worked completely fine with just the chiller," explained McCarry. "I think it's an important point to make: we didn't have all of our methodology figured out, we were still improving our knowledge, even as we were getting recognition as North American leaders in this field. But as innovative as we thought we were, our calculations had been off significantly – and yet the system still performed great."

The success of Library Square and its breakthrough underfloor air and ice storage systems quickly resulted in numerous design teams visiting the building from across North America.

We made a deliberate effort to make it more than just a repository of books. It's really a place for people. It's bustling, and people are very proud of it.

– Bing Thom, Vancouver architect and juror for the Library Square design competition[4]

Room for reading: Library patrons may use one of 1,200 reading galleries and arcades circling seven floors – with access to more than 1.5 million books, periodicals, photographs and recordings.

White Rock Operations Building

Something Old Made New Again

White Rock, British Columbia

There is more to the City of White Rock's new Operations Building than meets the eye. Much more. The combination of multiple green strategies within a single project was essential to achieving its LEED® Gold rating – the second LEED® Gold in Canada, and the first for new construction in the country.

The facility's breakthrough design incorporated several innovative approaches to energy reduction, with relatively short payback periods, as well as creative solid waste reduction and air quality solutions. The project is also exemplary for its emphasis on re-use, recycling and restoration: re-use of an existing structure; heavy use of reclaimed

Photo Courtesy of Busby Perkins + Will

Entering the Operations Centre: As designed by Busby Perkins+Will, the service/support facility for the City of White Rock is made up largely of recycled and locally-produced materials appropriate to the building's typology and the region.

Primary Team Members

Owner:
City of White Rock

Architect:
Busby Perkins+Will, Vancouver, BC
Project Team: V. Berg, P. Busby, D. Dove, S. Edwards, R. Maas, A. Malczyk, A. Slawinski, K. Wardle

Engineers:
Mechanical: Keen Engineering, Vancouver, BC – B. McCarry (Engineer of Record)
Structural: Fast + Epp, Vancouver, BC
Electrical: Flagel Lewandowski, New Westminister, BC

Consultants:
Landscape Architecture: Wendy Grandin/Viewpoint Landscape Architects Ltd., Vancouver, BC
Indoor Air Quality: Pacific Environmental Consulting Services, North Vancouver, BC
Cost: Helyar & Associates, Toronto, ON

General Contractor:
KDS Construction, Ltd., Surrey, BC

According to First Nations' legend, White Rock's giant, 486-ton white boulder marks the spot where the Transformers, emissaries of the Sagalie Tyee, the Creator, once passed.

and recycled construction materials; and extensive restoration of the site's landscaping and stormwater scheme.

It is fitting, too, that one of the most distinguishing features of the new Operations building is water. Located south of Vancouver along the British Columbia coast, White Rock is a scenic oceanside residential community clustered around the warm, shallow waters of Semiahmoo Bay. Only minutes from the Canada/U.S. border, this city of more than 19,700 people has also been a holiday destination for decades – renowned for spectacular views of the ocean, magnificent sunsets and wide sandy beaches, and its most famous landmark: a massive white boulder weighing 486 tons.

As one of the sunniest locales in the Lower Mainland, the area's moderate climate is also very wet, with an average annual rainfall of forty inches. This climatic feature provided a unique opportunity for the city's new operations center and site to demonstrate several stormwater management and water conservation best practices.

The project's beginnings began modestly enough. The City of White Rock needed more space for its Engineering and Operations staff who were being housed in a series of trailers and a small, outdated wooden-frame building in the City Works Yard. In 2001, the city requested proposals from architectural firms in an invitation-only competition to reinvent the Operations building. White Rock's mandate was clear: that the facility should be designed as environmentally sustainable as possible, in accordance with its own municipal policy of promoting green strategies in all current developments and planning.

Later that year, the firm of Busby Perkins+Will of Vancouver was selected to design the new facility – and their team included KEEN Engineering. The city's RFP was clear that the design team deliver a truly "green" building. And key to meeting that requirement meant an integrated design process from the outset, with the full design team contributing to all major design decisions.

Located In one of British Columbia's sunniest regions, the building design incorporated a 2.0 KW grid-tied solar array to generate more than 4 percent of its annual electricity needs.

Environmental Objectives

Expanding on the city's green mandate, City Engineer Greg Scott also outlined an aggressive goal to pursue LEED® Gold certification for the new facility. He was motivated, in part, by the increasing number of municipalities throughout the U.S. Pacific Northwest, particularly Seattle, where several LEED® Silver buildings were currently under construction. As the design evolved, it was also evident that the LEED® rating system would be useful in gauging the success of the many green strategies planned for the project.

Copyright © Jim Burns

We are water – the oceans flow through our veins... As air is a sacred gas, so is water a sacred liquid that links us to all the oceans of the world and ties us back in time to the very birthplace of all life.

– David Suzuki, *The Sacred Balance*

Project Program and Stats

Location:	White Rock, BC, Canada
Building Type:	Municipal Facility – New Construction
Size:	Site: 103,305 sq.ft.; Building: 6,545 sq.ft.
Stories:	Two-story building with basement component on the north end; one-story component with basement on the south end
Building Features:	Single building; two-pavilion structure
Site Issues:	Located on a hill overlooking a residential neighborhood, with old operations center building and buried tanks of a decommissioned water treatment plant
Completion Date:	May 2003
Rating:	LEED® Gold Certification – the second in Canada, and the first LEED® Gold for new construction
Cost:	$1.2 million (CAN)
Awards:	· Consulting Engineers of BC Award of Merit, 2004 · Architectural Institute of British Columbia "Lieutenant Governor of BC Medal" Award of Excellence, 2004 · Architectural Institute of British Columbia Innovation Award, 2004 · AIA/COTE Top Ten Green Projects, 2004

In setting the project's direction to ensure that the Operations Centre would be a "deep green" development, the city and project team also agreed the solid waste strategy should focus heavily on re-use and reduction. Instead of completely demolishing the old structure and re-pouring foundations, the new center would be located on the site of the original building and utilize the existing foundation of an old pump house and the buried tank walls of the decommissioned sewage treatment plant. In addition, a concerted effort would be made to eliminate waste related to construction, with recycling bins set up on the site for materials such as wood, metal, glass, asphalt and concrete.

Design of the Centre's two-pavilion structure made use of buried tank walls/concrete clarifiers from a decommissioned wastewater treatment plant for one foundation and the existing basement of the old treatment plant control building for the other.

Design Overview

Pre-design of the new White Rock Operations Centre concentrated on its primary function as a service and support facility for nine full-time staff and the eighteen field staff who are typically present two to three hours per day – plus approximately fifty visitors per week. The site is also used as a base for servicing and cleaning the city's fleet of civic utility vehicles and street-washing vehicles.

The entire project team was involved in the integrated design process, enabling additional efficiencies between team members and maximizing overall cost effectiveness during its green design and construction phases.

The City Works Yard is located on a hill overlooking a predominantly single-family neighborhood, so the new building was designed to fit into this residential context. It was split into two separate pavilions: a two-story component on the north end of the site, and a one-story building on the south end. The north building was designed to utilize the existing basement of the old sewage treatment plant control building and houses departmental elements used only periodically; namely, field crew facilities, changing rooms, a first-aid room, and meeting and lunch rooms. The south building features offices and an administration area. Its roof was designed partially as a roof deck, with the remainder featuring a "green" planted roof.

Designed as a post-disaster structure, building materials were selected for durability and appropriateness to both the building typology and the region and its climate. The new wood-frame building includes reclaimed heavy timbers and roof decking. A trellis screening the west-facing office pavilion is constructed of recycled telephone poles. Use of recycled materials in the design also included rubber flooring, insulation, gypsum board and carpet. All new concrete, including the cementious wall panels, contained an average of 40 percent fly ash. In total, 8 percent of the building products contained either post-consumer waste, post-industrial waste, or a combination of both.

The architects also made a conscious effort to select locally-produced materials, where possible, in order to reduce the energy consumed in transporting building materials to the construction site. All cladding, for example, was locally harvested and milled West Coast cedar. As a result, more than 31 percent of the materials were procured from manufacturers within a five hundred-mile radius of the site.

Building Performance (metrics)

The variety of sustainable strategies employed at White Rock is expected to reduce site water use by approximately 90 percent (2 million liters/year) and building and process water usage by at least 30 percent. It is also anticipated that approximately 455,000 imperial gallons of potable water will be conserved annually. Estimated water projections are:

- Indoor potable water use: 56,800 gal/yr (215,000 liters/yr)
- Potable water use per occupant: 2,670 gal/person/yr (10,100 liters/person/yr)
- Potable water use per unit area: 8 gal/sq ft (353 liters/sq meter)

Through these combined efforts, a reduction in energy consumption is expected to be 60 percent better than the Canadian Model National Energy Code for buildings (and 50 percent better than ASHRAE 90.1 - 1999). Energy-efficient strategies for mechanical and electrical systems are also expected to save approximately $5,000 (CAN) per year. The additional cost for high-efficiency systems will be paid for in just eleven years, after which time these strategies will continue to provide operational savings to the client.

MEP Design Process

With temperatures in White Rock averaging 23°C (73.4°F) during the summer and 6°C (42.8°F) in winter months, KEEN and the architect collaborated on designs that allow the building to maintain a very comfortable environment with passive systems. As such, the mechanical systems are minimal; where used, they are exposed for easy access and maintenance.

Because the building form was somewhat predetermined (reflecting the plans and geometry of the existing buried tank walls on site), designers specified glazing and solar protection treatments that would respond to the structure's orientation relative to the daily and annual solar path. As a result, north elevations have less than 5 percent glass, and office areas (to the south) have walls with 50 percent glass. East-facing windows are protected by a large roof overhang and deciduous trees. South windows are protected by a roof overhang and solar shades. West-facing windows are screened using the trellis that supports vines.

The building's office area was designed as a narrow bar – a small dimension to the south and larger dimensions east and west – which accentuates cross-ventilation. As a result, all occupants sit adjacent to operable windows for natural light and ventilation, and cooling loads are reduced (with no mechanical air conditioning). A green roof on the top of the office further reduces heat gain.

Left: The planted green roof over top of the administrative offices has dramatically reduced stormwater runoff.

Above: Aluminum sunshades (south elevation).

The two-story component of the building features a white, low albedo roof to reflect heat. Inside, KEEN designed a heat recovery ventilation system that "boosts" airflow, heating or cooling as needed during the day in transient spaces such as when field staff use the changing rooms in the morning, at lunch and at the end of the work day. In effect, the ventilators rise up to provide the necessary airflow requirements and then drop back down into "sleep" mode.

Old ideas can sometimes use new buildings. New ideas must use old buildings.

– Jane Jacobs, urban planner

The building design also features active solar strategies to conserve energy and minimize air emissions. Systems include solar hot water

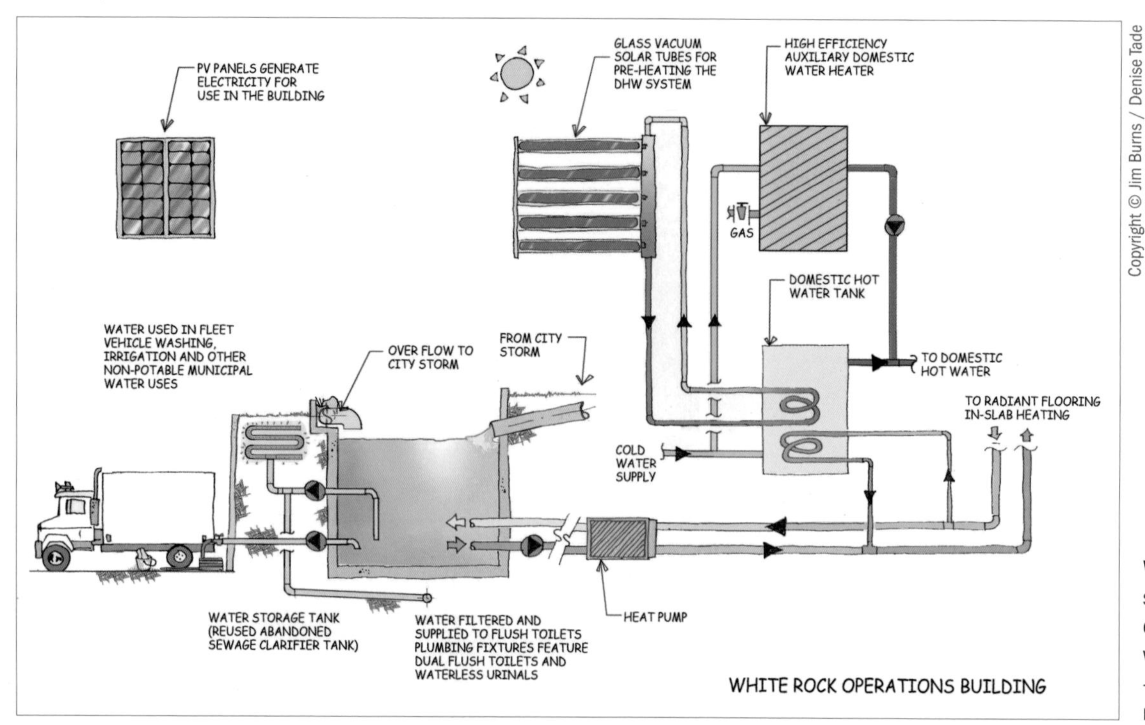

Water, water everywhere: KEEN's concept schematic reflected the huge potential of stormwater collection and re-use for a wide range of applications at the facility – lowering potable water consumption by nearly 500,000 gallons a year.

tubes, radiant in-floor heating, and a solar-panel array demonstration project that generates approximately 5 percent of the building's energy requirements.

The project's water conservation and stormwater strategies were equally innovative. The abandoned concrete storage tank of the treatment plant presented a unique opportunity for stormwater storage and re-use. In preparation for the redevelopment project, the city's engineering group came up with the idea of redirecting some existing storm drainage lines to discharge through the large, round 108,000 gallon storage tank. Given the region's high annual rainfall, that idea represented approximately 3.1 million gallons available for irrigation, flushing toilets, washing municipal vehicles and filling street-washing vehicles. KEEN recommended the addition of a water-source heat pump, which then utilizes water in the tank to provide heat to the building.

And finally, security/dependability of the energy supply for the new center was addressed in several ways. The building utilizes a number of low-tech strategies and technologies, reducing its dependency on energy from off-site sources. The site is serviced by an on-site emergency generator, so the building will be able to function independent of the grid. The City of White Rock bought green power certificates for all of the remaining annual energy requirements.

Indoor Environmental Quality

In meeting LEED® Gold criteria, the architect specified that all paints, sealants, adhesives, carpets and composite woods emit very low or no VOCs to minimize pollutants within the interior environment. Additionally, the passive design of the building relies strongly on natural ventilation for the comfort and viability of interior spaces and eliminates the need for conventional air conditioning. The building's office spaces were designed to include large expanses of glazing, with all works paces adjacent to operable windows allowing occupants to control their environment and gain outside views.

Water and Waste Design

Water conservation was a major design goal for the new city building. Highly efficient, water-conserving plumbing fixtures and waterless urinals are used throughout the complex to reduce water consumption. Drought-tolerant plantings minimize the need for irrigation. The green sod roof over the office wing reduces runoff from impermeable surfaces on site. The gravel parking lot allows infiltration of water into the ground.

Stormwater management became an equally significant feature of the operations site and building. One decommissioned sewage treatment tank was excavated and upgraded for use as a holding tank for all stormwater diverted from streets up the hill, north of the building site. Any overflow from the tank is then moved back to the city storm drainage system. Water collected from the tank is used to service all toilets for sewage conveyance, provide supplemental heat to the building, and irrigate landscaping. Another significant water use, estimated at 150,000 gallons per year, is for washing down, on a weekly basis, all city vehicles, which are white in color. Furthermore, the City of White Rock operates street washing vehicles, saving another 260,000 gallons of potable water per year.

Heating and Cooling

Sustainable heating and cooling design for the complex included several key passive and active strategies. Roof overhangs and plantings around the complex reduce solar gain, as does solar shading on the south side through use of exterior louvers, awnings and trellises. Increased insulation and reduced openings on the north face also minimize heat loss. Solar hot water tubes serve as the building's primary source of heat, supplemented by a high-efficiency boiler.

Designed as a narrow bar, White Rock's office area places occupants next to operable windows for natural light and cross-ventilation.

Going green delivers an excellent return – environmentally, economically and socially.

– Greg Scott, City Engineer, City of White Rock

Solar hot water tubes - along with radiant in-floor systems - serve as the building's primary source of heat.

In addition, thermal energy from stormwater diverted to the outdoor water storage tank (previously the old treatment plant clarifiers) was used to augment heating and cooling in the facility's office areas. In essence, KEEN designed a water-source heat pump system, utilizing water from the in-ground tank and passing it through a re-heat coil to heat or cool ventilation air, according to the season.

Blair McCarry, KEEN's lead designer on the project, had been intrigued by the potential of using lower temperatures – approximately 110° F – for radiant-floor heating and cooling:

> The major, occupied office areas did not require any air conditioning last summer because of good cross ventilation and operable windows. But when they do get a very hot day, the staff simply turns on the heat pump for some radiant floor cooling. During the winter, the radiant floor heating ensures comfort using low-level supply air for ventilation effectiveness, and heat recovery for energy effectiveness. It works really well.

Lighting & Daylighting

The project's daylighting strategy was simple: locate frequently-used areas on the south side of the building, with large interior windows to increase daylighting penetration; and light shelves and shading devices that redirect outside light and control glare. KEEN and the architect gave careful consideration to visual comfort. They selected interior and exterior glazing to maximize daylight transmission and an open-floor plan design that permits daylight to penetrate into the interior as far as possible.

The office area also includes occupancy sensors to minimize use of electricity in unoccupied spaces. Each workstation is furnished with high quality task lighting, which allows overall ambient light levels to be reduced and gives occupants greater control of their interior work environment.

Site

Because it was previously-developed land, the client wanted to emphasis low-impact siting for the new building. An earlier site study performed by other architects had recommended the city avoid the portion of the site where the sewage treatment plant was located, due to "unknowns related to this area." Busby, KEEN and the project team saw it differently, believing the abandoned tanks would be an asset as part of water conservation, stormwater and even thermal energy strategies.

Additionally, the landscape design included dramatic increases in open space around the building through reduction of paved areas on the site. As a result, 4758 square feet of parking surface was replaced with grass/gravel pave and another 4,050

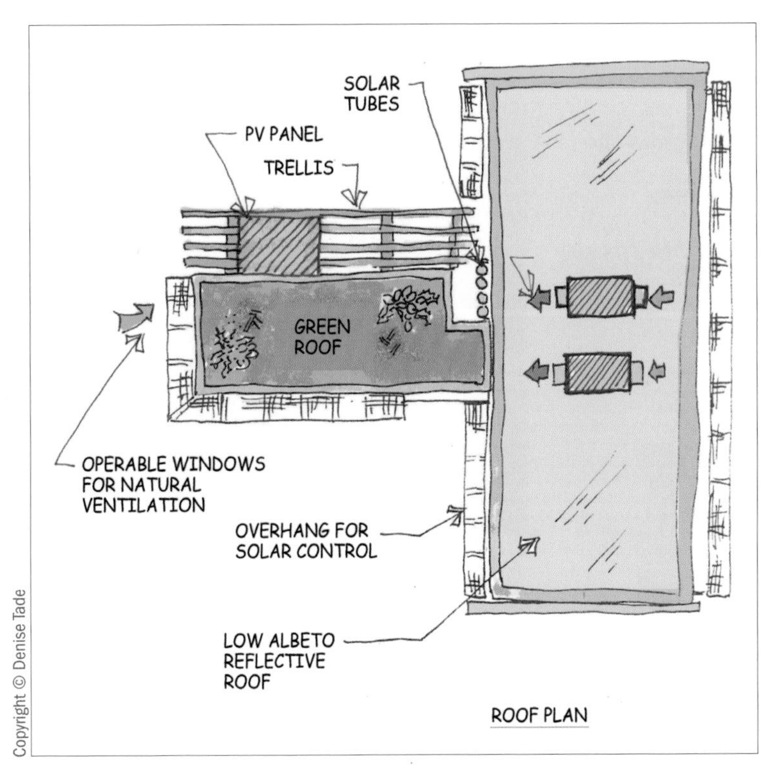

Solar control: KEEN's recommendations for White Rock's sunny climate included large roof overhangs, exterior louvers, awnings and trellises (south elevation), a PV array, a green roof and low albedo reflective roof.

square feet of paving was returned to nature. Planting throughout the site makes use of local drought-tolerant plantings, some of which were relocated from another site redevelopment where they would otherwise have been destroyed instead of helping to lower maintenance needs. The Yard is well screened from the street by mature landscaping, and the siting of the new building minimizes any negative effects on neighbors' views.

Renewable Energy

Active and passive solar energy systems, as well as thermal energy via heat exchangers and the water storage tank, were vital elements of the new center. The City of White Rock also purchased BC Hydro Green Power Certificates to augment its electricity needs through the use of solar tubes to provide the base radiant heating for the building.

The facility became the first building in Canada to achieve LEED® Gold status using solar power for on-site renewable energy. To gain this credit, a building must generate 5 percent – approximately 2,900 kWh per year – of its energy consumption from an on-site renewable energy resource. For this application in White Rock, the energy source is a 2.0 kW grid-tied solar array, designed using 16 Shell SP-130 solar modules and an 1800 watt inverter.

Engineering Process

Throughout the design process, the team debated decisions on systems and modeled options to ensure that all systems incorporated into the building were optimally suited to the site, the program and the client. The design team also consistently measured its sustainable strategies against the LEED® Rating System.

Verification of the design intent continued throughout the construction process. As part of LEED® certification, each credit had to be documented and many of these were dependent on the final execution of the design, not just the design itself. During construction, the contractor worked with the design team to ensure that the requirements and intent of each LEED® credit were met and that adequate documentation was provided. As such, the construction phase was very collaborative.

Solid waste reduction was an integral goal of the project as well. Specifically, 98 percent of all construction and demolition waste by weight of the existing office building was reused or recycled. This reuse/recycle component represented more than 467,670 tons diverted from the area landfill.

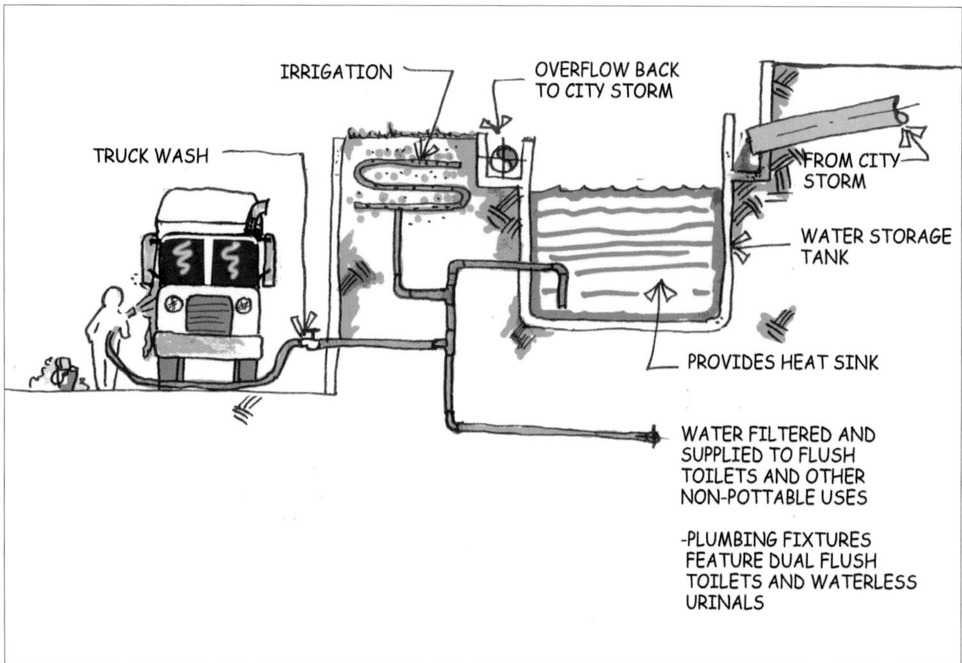

Stormwater is collected and diverted across the Operations Centre site into a 108,000-gallon holding tank – once a sewage treatment tank – then treated and available for toilet flushing, landscape irrigation and washing the city's fleet of service vehicles.

The client was equally instrumental in achieving the final "green" result for this building through the city's creative financing efforts. On behalf of the Canada/British Columbia Infrastructure Program, a joint federal and provincial fund that supports green infrastructure, the City of White Rock received $898,000 (CAN) in funding. Other sources of funding came from the federal CBIP (Commercial Building Incentive Program) incentive, which pays for additional design efforts required to assist the design team in achieving reduced energy consumption. The REDI Grant, sponsored by NRCan (Natural Resources Canada), also provided funding to the project for the study of a variety of design strategies that could improve the building's environmental performance through renewable energy solutions.

And finally, when construction was complete, KEEN and the team's other engineers carried out complete commissioning on the building to ensure that systems were operating according to the design intent.

C.K. Choi Building for the Institute of Asian Research

A Model of Possibilities

Vancouver, British Columbia

The University of British Columbia lies at the tip of a peninsula, west of Vancouver. In the northwest corner of the campus sits a striking building, with a series of five gracefully-curved metal roofs rising along its length, aimed towards the North Shore Mountains, reaching for the sky. The tinted purple color of its many-windowed brick façade contrasts with the towering coniferous trees – native cedars and Douglas firs – behind the building. This edifice is the C.K. Choi Building for the Institute of Asian Research. If ever a building had a heart and soul, this is it. A place of sun and shade, fresh air, water and earth, this sustainably-designed structure has become one of the most visited and studied green buildings of its era.

Through well-conceived site planning and design, effective energy, water, and waste management, and control of thermal comfort and indoor air quality, the C.K. Choi Building has become a model for sustainable architecture among educational institutions.

Primary Team Members

Owner:
University of British Columbia – F. Pagani (Associate Director, Project Development for Campus Planning and Development); B. Murfitt, J. Anderson (Project Managers)

Architect:
Matsuzaki Wright Architects Inc., Vancouver, BC
Project Team: E. Matsuzaki (Principal-in-Charge), J. Perdue (Project Architect)

Engineers:
Mechanical: KEEN Engineering, Vancouver, BC – K. Hydes (Engineer of Record), J. Frost (Project Engineer)
Structural: Read Jones Christoffersen Ltd., Vancouver, BC
Electrical: Robert Freundlich Associates, Vancouver, BC

Consultants:
Landscape Architect: Cornelia Hahn Oberlander, Vancouver, BC
Landscape Contractor: North by Northwest Landscape, Vancouver, BC
Sustainable Consultant: BNIM Architects, Bob Berkebile, Kansas City, MO

General Contractor:
Country West Construction Ltd., Abbotsford, BC

Never doubt that a small group of thoughtful, committed citizens can change the world. Indeed, it is the only thing that ever has.

– Margaret Mead

Creating the C.K. Choi Building began in earnest in March 1993 with a day-and-a-half of inspiration. The University wanted to launch the project with a "common vision" workshop, with all key stakeholders in attendance to share joint goals and set objectives as a group. Led by Bob Berkebile, the Kansas City-based architect and founder of the AIA's Committee on the Environment, the workshop was designed as a critical first step in encouraging early collaboration among the consultants and disciplines and in drawing consensus from all parties.

A total of twenty-five participants gathered in UBC's Asian Centre next to the new building site. They included building users from five academic departments, the dean in charge of the facility, representatives from the University's planning office and physical plant and building maintenance departments, as well as the entire team of architects, structural, mechanical and electrical engineers, and landscape architects.

"It was like somebody had switched on a light," remembered Kevin Hydes. "Bob's very first request of everyone was to say something personal about themselves and what they believed in. I had never actually experienced anything like it before, and probably neither had anyone else in that room. Those two days allowed us to express ourselves in a genuine way, and it created a sense of reverence for the meeting.

"The way I see it is, we were the mechanical engineers who designed a building without a mechanical system," he said. "That's the best way to describe it, the building became the system. That was a big shift for us, because, traditionally, our fees were tied to the size of the mechanical system. In this meeting, in this environment, that didn't even become a consideration. It was obvious that we could design it most appropriately as a group and in a way we really believed to be the right way."

Just as important was an unwavering commitment from the University to the project, conceived during a period of rapid growth on the seventy-five year-old campus. In the midst of a ten-year, $500 million building expansion program, UBC's fundraising campaign would become the most successful in Canadian history. Funding for the building came from a number of major donors in Canada and Asia, including a generous donation from Dr. Cheung-Kok Choi and family and matching funds from the Government of British Columbia.

The University of British Columbia's C.K. Choi Building: Creating a new paradigm for green.

Zhi (Wisdom) can see things not yet materialized. It covers all matters. It enables one to distinguish between right and wrong and to respond to a situation in an appropriate and moral way. The ideal of zhi is wise decision making.

– Confucian Analects
from the Stone Garden at the C.K. Choi Building

Project Program and Stats

Location:	Vancouver, BC, Canada
Building Type:	University Office Building - New Construction
Size:	Building is 34,400 sq.ft.; lot is 18,000 sq.ft.
Stories:	Three stories
Building Features:	Narrow profile - north/south orientation; five distinct towers/atria topped with curved roofs
Site Issues:	Urban campus, previously developed land, with second- percent of the site
Completion Date:	April 1996
Cost:	$4.5 million (CAN)
Awards:	· AIA/COTE Top Ten Green Projects, 2000 · Architectural Institute of British Columbia "Lieutenant Governor of BC Medal" Award of Excellence, (Matsuzaki Wright Architects), 1998 · Building Owners and Managers Association's Earth Award, 1996 · Consulting Engineers of BC Award of Merit, 1996 · BC Hydro Power Smart Award of Excellence, 1996

Initially, the University's planning office had no green agenda, nor did its procurement rules contain environmental guidelines. Freda Pagani, then Associate Director, Project Development for Campus Planning and Development at UBC, had made a personal commitment in late 1992 to pursue greater environmental consciousness with all new buildings constructed on campus. She saw the new Choi facility as an opportunity to demonstrate sustainable design and she was supported by several key faculty members. Although the selection of an architect would be based primarily on experience, the ideal candidate also needed to be highly enthusiastic about green buildings. In its interview, Matsuzaki Wright Architects of Vancouver met both criteria.

On the second day of the visioning workshop, Berkebile had skillfully translated the group's statement of objectives for the facility into a shared vision:

- To establish the benchmark for state-of-the-art collaborative research
- To express the character of the cultures participating in the research
- To set new standards for sustainable design, construction and operation

The University's master plan of the campus showed the proposed site as an area of mostly second-growth forest, with the new building represented as a square-shaped dotted line. The plan also showed an existing strip of asphalt parking, sixty feet wide by three-hundred feet long, near the north entrance to the UBC campus. Over the next three years, the building gradually took shape - three stories high and more than 30,000 square feet of space with a north/south orientation.

Eva Matsuzaki, the lead architect, is convinced that maintaining that shared vision as a group was vital to the project's success:

> Often you set out goals and then there are a lot of compromises.

By preserving the native fringe of cedar and Douglas fir on the site (western elevation), designers have given building occupants a sense of a "clearing in the forest."

> But it was just a wonderfully dedicated team, and each one of them was absolutely committed: the landscape architect, the electrical engineer, the structural engineer, the mechanical engineer. So when I look back, I'd say the team had the passion at a very emotional level that lasted for three years, from the project's start to when it was built.

The C.K. Choi Building for the Institute of Asian Research officially opened on October 7, 1996, with principal donor C.K. Choi, his son David Choi, and other members of the Choi family joining University dignitaries at the opening ceremonies. Today, the building is more than innovative architecture. It reflects a new approach to research on Asia and it is a symbol of sharing between East and West. The Institute is a building of many countries, housing academic research offices and seminar rooms for five research centers that focus on China, Japan, Korea, Southeast Asia, India and South Asia – and increasingly, as a center for visiting scholars, international conferences, art exhibits and cultural performances.

Environmental Objectives

From the time of its first planning meetings in 1993, the C.K. Choi Building was designed to be a "benchmark in sustainable design." Environmental objectives were identified in four key target areas: Reducing Environmental Impact and Consumption; Minimizing the Energy Needed for Construction; Lowering Operating Energy Over Time; and Creating a Liveable Working Space. The project team was challenged to address industry standards at every stage of the process and jointly establish criteria for the design, construction and operation of the building.

Specific targets included:

- 50 percent less water use than normal
- No sewer connection
- 50 percent re-used/recycled materials
- 50 percent recyclable materials
- Reduction of energy use below ASHRAE 90.1 levels by 35 percent
- Lighting for less than five watts per square meter [0.5 W/ft^2]

To achieve these targets, the team identified several key architectural design and MEP strategies. To reduce environmental impact/ consumption, the team would explore several water-conserving and waterless measures – enabling the building to minimize and even eliminate the need for a connection to UBC's sanitary infrastructure. To minimize energy usage in construction, it was estimated that more than 50 percent of the total materials incorporated into the building could be reused or recycled. To lower operating energy and expenses, the building would rely heavily on daylight to eliminate conventional lighting and on natural ventilation instead of ducted air systems.

Design Overview

Architects Eva and Kyoshi Matsuzaki clearly defined the C.K. Choi Building by locating the five academic departments in five distinct towers anchored by a low central horizontal block. Their aim was to give a physical presence, both on the interior and exterior, for each research center while maintaining a unified look with no single center or culture dominating. Each of these Asian studies departments had been located previously on different parts of the sprawling UBC campus. The University's consolidation plan also called for shared resources in research spaces while maintaining a level of autonomy

Left: Recycled materials were used extensively throughout the building, including its brick façade – pavers reclaimed from the streets of Vancouver.

Above: The 30,000-square-foot Institute was designed to feature five towers, each housing a distinct Asian study area focused on China, Japan, Korea, Southeast Asia and India and South Asia.

in administrative activities between the departments. To foster interaction between departmental staff, a central multi-purpose space was created on the ground level.

To give the building its Asian feel, curving galvanized steel roofs top each of the five atria, rising up beyond the three-story height and featuring north-facing windows – natural light wells that help illuminate second- and third-floor areas. Because of the building's narrow, linear configuration, it was possible to locate the majority of workstations beside large exterior windows, allowing occupants to

The Eastern soil embodies the ancient traditions while the Western way embraces the latest technology. As the five continents gather more closely, the whole world beats to the rhythm of one heart.

– Dr. Cheung-Kok Choi

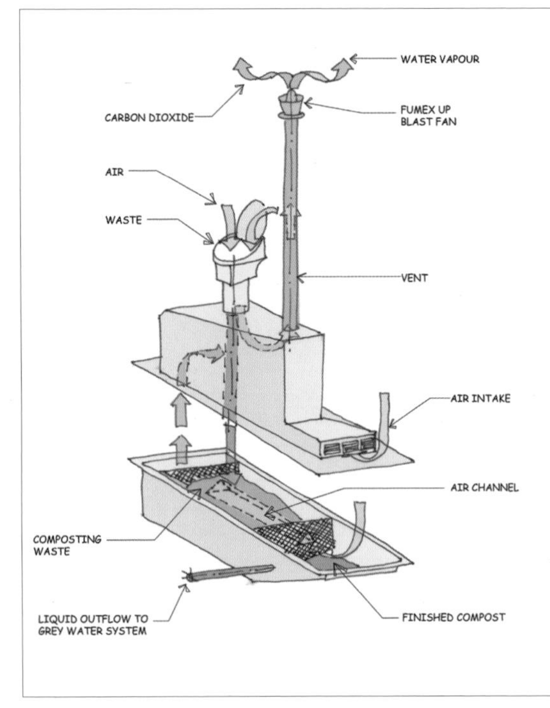

HOW DOES IS WORK?
ORGANIC CARBON FOUND IN THE WASTE AND BULKING AGENT ALONG WITH THE NITROGEN FROM UREA PROMOTE THE GROWTH OF AEROBIC BACTERIA. THIS BACTERIA BREAKS DOWN THE WASTE INTO WATER VAPOR, CO_2 SAFE COMPOST, AND A LIQUID END PRODUCT, RED WIGGLER WORMS ASSIST IN THE BREAKDOWN OF WASTE.

CHARACTERISTICS OF COMPOST
AT THE END OF A YEAR'S USE OF DRY PRODUCT IS PRODUCED WHICH CAN BE USED AS NUTRIENT RICH POTTING SOIL. 30,000 USES PER YEAR WILL PRODUCE APPROXIMATELY 4 CUBIC FEET OF USABLE COMPOST.

CHARACTERISTICS OF LIQUID END PRODUCT
A NITROGEN RICH LIQUID END PRODUCT IS PRODUCED. THIS CAN BE DILUTED 1:10 WITH WATER FOR USE AS FERTILIZER.

100 USES PER DAY PRODUCES 3-5 GALLONS OF LIQUID.

ARE THE BY PRODUCTS SAFE?
YES-THE VANCOUVER DEPARTMENT OF HEALTH HAS REVIEWED THE TEST DATA FOR BOTH END PRODUCTS AND HAS AGREED TO THE USE OF THE SYSTEM. THEY HAVE REQUESTED THAT A CONNECTION FORM THE CAMPUS SANITARY LINE BE BROUGHT TO THE SITE AS AN EMERGENCY BACK UP.

WHAT IS THE FAN FUNCTION?
TO AID IN THE REMOVAL OF ODORS, CO_2, WATER VAPOR, AND ALLOWS FOR THE INTRODUCTION OF O_2 TO HELP BREAKDOWN WASTE.

WHAT IF THE FAN FAILS?
THE EXHAUST SYSTEM CONSISTS OF DUTY AND STANDBY FANS. AN ALARM MESSAGE ON THE DDC WILL BE ISSUED AND THE STANDBY FAN WILL START.

WHAT HAPPENS DURING LONG PERIODS WHEN THE TOILETS ARE NOT USED?
THE COMPOSTING PROCESS CONTINUES. THE FAN CONTINUES TO RUN. A MISTING SYSTEM WILL BE PROVIDED TO ENSURE THAT BACTERIAL ACTIVITY CONTINUES.

With no cost-effective way to connect the building to UBC's aging sanitary system, KEEN suggested composting toilets as an alternative – the first large-scale commercial installation in North America. The anaerobic composting process turns the solid waste into topsoil and saves the university an estimated 100,000 gallons of water annually.

control daylight and air changes.

The architects aimed to control interior finishing as rigorously as the air and light requirements. Materials were chosen for their lack of "off-gassing". No adhesives were used to lay natural-fiber carpets, the paint is acrylic, and photocopy machine areas were designed with extra venting to combat toner emissions.

Two distinctive visual features of the C.K. Choi Building carry on local history in a new location: red brick cladding on the outside and heavy timber structural beams on the inside. The reclaimed brick was now entering its "third life": the bricks were first used as ballast on ships coming over from Asia in the 19th Century and then used to pave the streets of Vancouver's Gastown District in the early 20th Century. Architects contacted demolition companies to learn about the varying quality of bricks in old buildings and determined that street pavers would be the most durable as an exterior, non-structural façade.

The heavy timbers were salvaged from the old Amouries building directly across the street, a 1930s campus structure slated for demolition. The large, ungraded wooden beams initially posed a challenge when the timber grader rejected most of the wood as unusable – it had "checks" (cracks) that could affect the beams' structural integrity. Later, the structural engineer was able to certify the beams' reuse in the building by eliminating the weak end-portions.

The architectural design called for the beams to be left exposed, creating an aesthetic juxtaposition to the engineered window wall and smooth concrete floor.

Other reused and recycled building materials were incorporated into the building; namely, the main interior stair handrail, atrium guardrails, all doors, sinks, toilet accessories and some electrical conduits. Other building components such as reinforcing steel, structural steel, insulation, plywood, drywall, framework timbers, doors and frames, ceramic tile, and even the polished concrete floors have recycled content. In addition, decorative finishes to functional materials, such as suspended ceilings, paint to the wiring trays and exposed pipes, floor finishes to concrete slabs, and cabinetwork were eliminated wherever possible.

Building Performance

The C.K. Choi Building was completed at US$95 per square foot, roughly the same budget as any other building on the UBC campus, even though the design and construction team discovered that using salvaged materials often required additional research and testing costs.

In this building, overall energy use (steam and electricity combined) is about 23 percent less than the ASHRAE 90.1 prototype building and 40 percent less than average building standards – with annual savings totaling about 423 GJ [401,000 Mbtu]. Metering also showed that actual electricity consumption is 28 percent below the design estimate and 69 percent lower than a similar ASHRAE 90.1 prototype building. Annual electricity consumption is approximately 101,000 kWh, compared to the design estimate of 142,500 kWh, and 334,000 kWh for the ASHRAE 90.1 prototype.

Total electrical savings are 191,603 kWh per year. For this achievement, BC Hydro provided a $44,121 incentive to the owner under the New Building Design Program.

Extensive tree cover behind the C.K. Choi Building was retained to enhance natural solar shading.

KEEN took advantage of the building's five tower atria to bring daylight into office spaces and utilize internal stack effects to promote natural ventilation.

I do not put my faith in any new institution, but in the individuals all over the world who think clearly, feel nobly, and act rightly, thus becoming the channels of moral truth.

– Rabindranath Tagore, Nobel Laureate

MEP Design Process

UBC's campus planners and engineers had often selected KEEN in the past for the firm's proven expertise with chillers and other big systems. Hydes and other individuals were currently working on more than twenty buildings across the fifteen thousand student campus. This time, however, KEEN's engineers knew they would have to abandon many of their traditional mechanical design concepts in order to achieve sustainability goals required of the Choi building. Hydes recalled how his first impressions of the proposed site fueled his design thinking toward this new project:

> When we saw this amazing stand of trees over 70 percent of the site and the strip of parking along the road, we realized the building shouldn't have any bigger footprint than the existing parking lot. Then we stacked three stories on this sliver of land, knowing that 30-foot-wide buildings in this climate could be naturally ventilated. There was precedent. So we drew a diagram with ideas of what the building could be – showing the stack ventilation, natural ventilation, daylight ideas – and took it to be interviewed as the mechanical engineer for the project. Certainly, we take no credit for the architecture – absolutely not. But it's interesting how close that diagram is to the finished building.

The dynamic building form capitalized on the benefits of the different climatic conditions at each orientation. Retaining the stand of trees along the west elevation would allow for natural solar shading and assist in removing CO_2 from the air. The narrow profile of the building would promote cross-ventilation, while its five tall atria would enhance internal stack effects to promote natural ventilation. As a result, the building's occupants could enjoy natural daylight and 100 percent fresh air year-round, further reinforcing the office as a comfortable place to work for forty hours per week for all three hundred occupants.

As the project progressed, KEEN engineers were increasingly asked questions they had never been asked before on their conventionally-

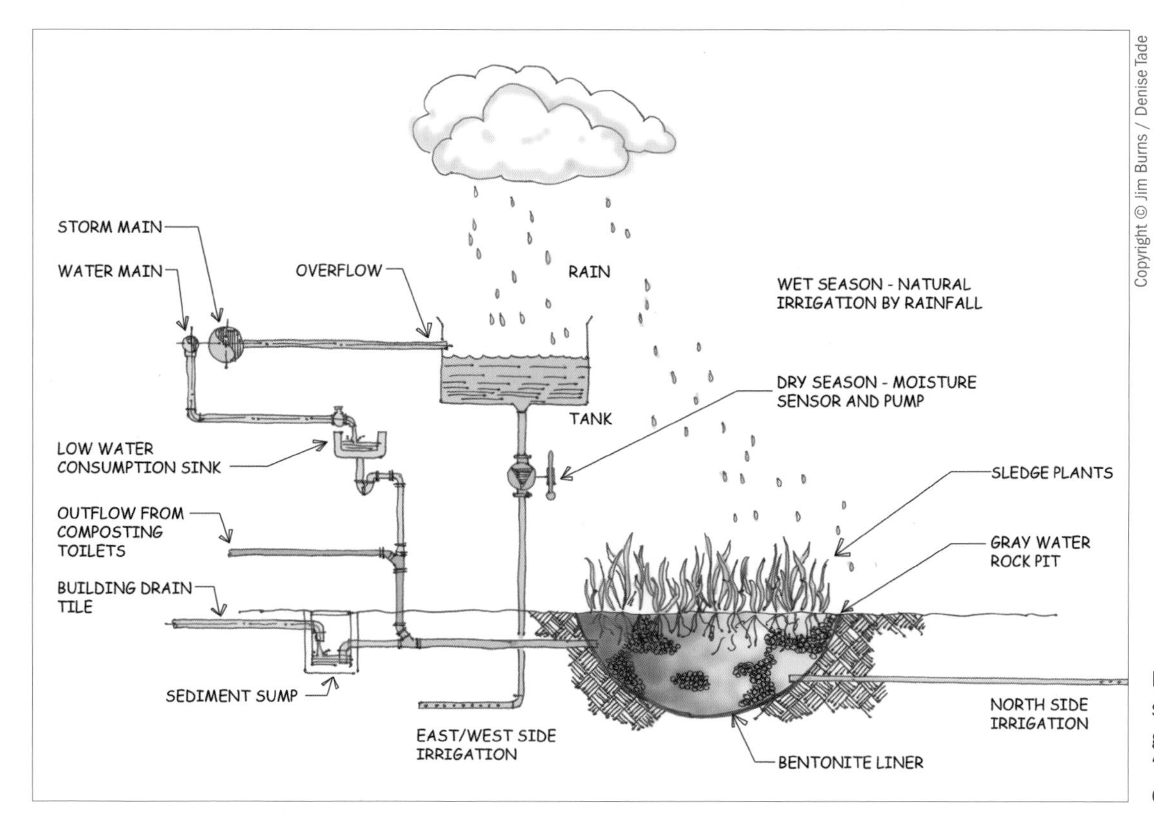

KEEN's concept for a wastewater treatment system at C.K. Choi – combining rainwater, greywater from sinks and composting "tea", then filtered through a subsurface constructed wetland for use in irrigation.

designed systems. They needed answers - anything they could find on daylighting and natural ventilation strategies to satisfy the code requirements. That search for answers meant referencing buildings of a similar proportion and in a similar location that had performed well over time. Other than empirical data, Hydes was unable to produce any calculations to satisfy the code. Some validation came from old design manuals. Some came from a recent U.K. design manual, "Natural Ventilation for Office Buildings." The twenty page manual provided intriguing – and accurate – data, but it did not match up precisely with ASHRAE nor with ranking Canadian standards.

Unquestionably, the design decision to install flushless, composting toilets in the C.K. Choi Building piqued a great deal of interest. Completely disconnected from the campus sewer, this waterless system has enabled the building to save fifteen hundred gallons of potable water per day.

Other key energy-saving design elements of the building included: high performance walls and window glazing; careful attention to detailing and construction methodology to minimize heat loss through thermal breaks; use of stored rainwater for irrigation; and use of waste heat from a nearby steam vault to pre-heat domestic hot water. Also, underutilized power from the adjacent Asian Centre was available so that new electrical service or transformers were not necessary.

Indoor Environmental Quality

Instead of a traditional ducted air system, the C.K. Choi Building was designed to rely on natural ventilation through operable windows, high intake louvers and fixed air-grilles below the windows to deliver 20 cfm per person of fresh outside air. This natural ventilation strategy provides continuous fresh air and ensures that carbon dioxide is kept to a minimum in occupied spaces. The cool air rises as it warms and exits the building through louvres high in the atria. Being continually flushed, the office space does not experience the peaks and valleys in indoor air quality often found with traditional mechanized systems that operate with reduced air changes during low occupancy hours.

The office layout and detailing also promoted effective air circulation. All workstations are located close to operable windows for health and comfort. Small localized fans provide additional circulation if the ambient temperature becomes too warm. In winter, the building uses trickle ventilators and hot water baseboard heaters which temper the air below the windows. In summer, occupants control their own ventilation using the operable windows.

Designers followed three additional strategies to ensure good indoor air quality: carpet was laid without adhesives; millwork was constructed from formaldehyde-free boards; and all finishes were solvent-free/low-VOC products. Construction sequencing was specified to ensure flushing of the building during drywall installation and finishing, painting and caulking, and carpet installation. Additionally, copy machine areas were situated to vent directly to the outside.

Water and Waste Design

Significant water savings have been realized through a series of design features. City water is used only for the low-flow lavatory faucets (spring-loaded to further reduce water waste) and kitchen sinks. Most importantly, the design did not require storm, steam, condensate or gas lines.

"We had been working on all these other buildings on campus, so we knew the University's steam lines and sewer lines were already at capacity," said Hydes. "To connect our building to the sanitary system, it would have been necessary to dig up the street and use bigger pipes, which was going to cost around $250,000. So we thought, why not design a building that didn't need any sewer or steam connections; therefore, we don't burden the system any further."

KEEN's solution: toilets that function without water – the first composting toilets on campus. Manufactured by Clivus Multrum, Inc. of Massachusetts, the Clivus composting toilet was invented in Sweden in 1939. The system is based on an aerobic composting process that is continually ventilated, reducing the volume of waste by 90 percent. The end products are liquid and a humus-like soil material that is rich in nitrogen and other useful elements.

Though the University became convinced of the merits of composting, persuading the Vancouver Health Department was more difficult – particularly since Vancouver's plumbing code did not address the process and a commercial building installation of this scale had never been done before in North America. KEEN was able to obtain case study data from the manufacturer. But the information was limited to single-story residences. However, by applying the same data points to a three-story, 30,000-square-foot office building, KEEN was eventually able to demonstrate similar performance and satisfy all health department concerns.

The composting process itself involves a system of five trays at the bottom of the toilets' fourteen inch stainless steel chutes. Maintenance staff throw in organic matter such as wood chips to aid the composting process, and red wiggler worms turn the solid waste into topsoil. Liquids from the system – "composting tea" – combine with greywater and rainwater runoff in a subsurface trench along the eastern, street-side of the building. Additional expertise on greywater systems was provided by KEEN's Toronto

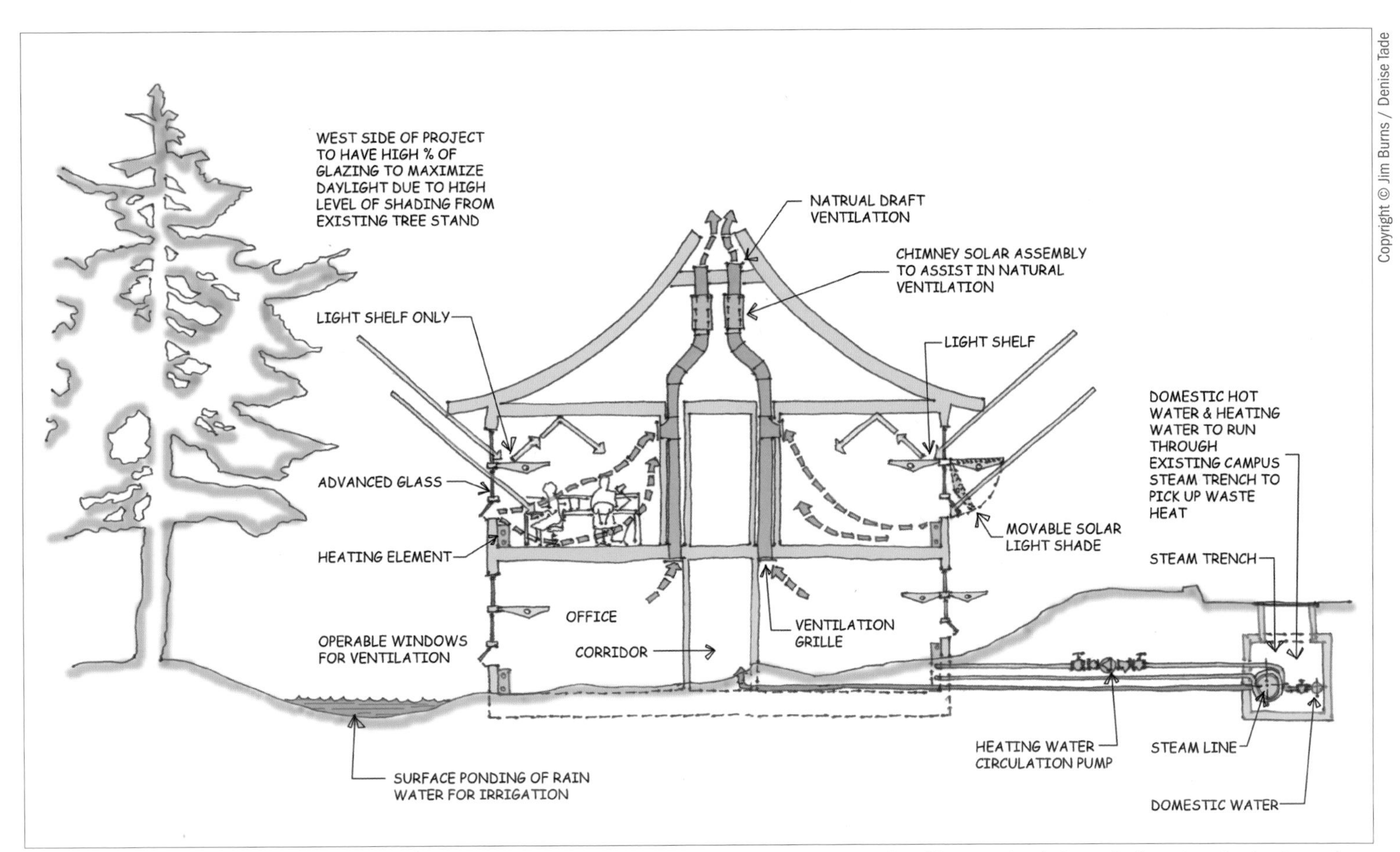

Concepts for C.K. Choi: "We drew the diagram, we actually showed the stack ventilation, natural ventilation, daylight ideas...and took that to be interviewed to be the engineer on the project." (Kevin Hydes).

office. The trench, lined with recycled PVC and filled with gravel, was created as a "biological marsh" using phragmite plant varieties such as reeds, sedges and irises. Micro-organisms on these marsh plant roots naturally purify the water as it is released gradually into an eight thousand gallon subsurface cistern; in turn, the greywater is re-used to irrigate site plantings during Vancouver's dry summer months.

The Clivus toilets, located on all three of the building's occupied levels, are estimated to save the University nearly 100,000 gallons of water per year. In addition, the Vancouver Health Department later tested the fecal coliform of the water and found it to contain 10 parts per 100 ml (swimming is allowed up to 200 parts, and stormwater varies from 1000 to 2000 parts).

Heating and Cooling

UBC's unique location on a peninsula creates favorable conditions for free cooling, particularly with the stand of trees as a source of cool, fresh air during the hot summer months. The heavy concrete structure provides the building with a high thermal mass and reduces the problem of diurnal temperature swings. Its mass cools down overnight, essentially storing the cooling energy.

Waste heat from the campus underground steam infrastructure is the primary source of space heating. Domestic hot water is produced in an electric hot water tank, which includes a preheat coil in a nearby steam manhole. The intent was to recover waste heat from the manhole but, in practice, the heat gain has been negligible.

In addition, low-e double-glazed windows and exterior insulation reduce thermal bridging. R-values suggested under ASHRAE 90.1 were exceeded for walls and ceilings, and the amount of heating equipment required was considerably less than for conventional buildings. Individual heating and cooling controls were specified as opposed to more costly DDC control systems.

Lighting & Daylighting

Inside the C.K. Choi Building, spaces are filled with daylight. The building's architecture maximizes natural daylighting through the use of extensive windows, light wells, and north and northeast exposure roof-mounted skylights. Wherever possible, natural light is introduced from two directions to soften and balance the lighting conditions. Atria admit daylight into the second-floor work spaces, as do higher ceilings on the ground floor. Designers also specified finishes with high-reflectance values, a high percentage of glazing on exterior walls, and high reflectance surfaces at grade around the building perimeter (instead of blacktop).

Daylight sensors and continuous dimming ballasts ensure that electric lighting is used only to supplement the available daylight. The interior lighting strategy incorporates suspended, high-efficiency luminaires with lower ambient lighting levels and task lights where appropriate. Occupancy sensors turn off artificial lights when not in use. The connected lighting load is approximately 0.9 watts/sf, so that power consumption for lighting is less than half that required for a comparable office building. The capital premium for the lighting strategies was calculated at $43,000.

Renewable Energy

South-facing roofs were designed to optimize direct solar gain and accommodate photovoltaic arrays as they become more economically feasible in the future.

Site

Situated between remnant forest on the west side and a busy roadway on the east, the C.K. Choi site was designed to retain and extend an arboreal sense so the building would occupy a "clearing in the forest." By locating the linear building footprint on the paved parking surface, the architects wanted occupants to experience the forest from all parts of the building.

A careful inventory was taken to assess the vigor and structure of all trees on the site. Existing trees were preserved, in part, for their capacity to remove carbon dioxide from the air and for the shade they provided on the west elevation. The building was also sited so that the tower atria would facilitate daylighting into work spaces through north-facing clerestory windows. Gingko trees, a variety known for their high capacity to absorb air pollutants, were planted along the street; their columnar shape meanwhile allows sunlight to penetrate the building.

Sustainable landscape design was another important element of site development. Landscapers installed a water-efficient, subsurface irrigation system based on the principle of xeriscaping. Native or indigenous plants were used around the building, particularly those requiring little maintenance. In addition, hard surface areas around the building are composed of interlocking pavers on a gravel base. These areas were sloped towards the forest and allow surface runoff to seep between the pavers and irrigate the soil and adjacent forest.

Engineering Process

Much of the C.K. Choi Building's success can be attributed to the integrated design process. All design consultants (the architects, structural, mechanical and electrical engineers, and landscape architects) met every Wednesday morning for two-to-three hour interdisciplinary design sessions through completion of working drawings.

Resulting innovations, such as greywater recycling systems, integration of recycled building materials and naturally-ventilated conference rooms, meant taking risks by deviating from standard practice. The proposal to use composting toilets in lieu of traditional water closets took literally hundreds of hours to research, present and, ultimately, gain approval.

Regular meetings between KEEN and the architect were also extremely important. "Really, the Matsuzakis were interested in just seeing what we were thinking and discussing," said Hydes. "So we'd come over periodically with little hand sketches of the ventilation or daylighting schemes. It was an outline developing."

Frequent interaction with the engineers on campus and the local health officers was also necessary and, at times, presented more hurdles to overcome. Yet, it also became essential to validating KEEN's breakthrough design ideas about water, air and comfort in the building.

"We forwarded a copy of the ASHRAE standard to the campus engineers with the words 'naturally ventilate' circled in red," said Hydes, "and we were told pretty much at every meeting that it's our liability. The reality is we were preparing ourselves to take a risk, there's no question about it. And honestly, those engineers were as important to the process as anybody, because they were always making sure we were performing our due diligence."

"What we were starting to explore, of course, is really understanding what the codes were based on in the first place. That was also the beginning, for us, of dispelling a lot of myths about ventilation."

To better understand the building's energy usage, the University agreed to participate in BC Hydro's Power Smart New Building Design Program. The project team then performed an energy study using DOE-2 simulation software to check and re-check the integrated design hypotheses and energy profile. The study confirmed that the 30,000-square-foot building would exceed ASHRAE 90.1, the advanced building protocol, by 57 percent and estimated its overall energy use at 9 percent less than ASHRAE 90.1.

Quality management checks were carried out at schematic design and design development, and at the 30 percent, 50 percent, 80 percent, 90 percent and 100 percent working drawing stages. Field reviews were conducted with the general contractor and subcontractors every two weeks during construction to monitor progress toward the project's sustainable goals. Cost estimates were done at various milestones to ensure cost reduction targets were realized.

A final, key piece of the project involved a series of intensive commissioning sessions held to familiarize users with the building's design and explain how to operate the various "low-tech" mechanical, electrical and control systems. Educating tenants was considered essential for optimal comfort within the structure. Users were also encouraged to provide feedback and assist in metering the building to monitor actual energy use and make necessary adjustments in operations.

Today, the University continues to lead tours of the C.K. Choi Building every other week, with interested visitors coming from all over the world. For her part, Freda Pagani is proud to have contributed to the creation of this landmark building, and she is astonished at the impression it continues to make on people:

> To me, it's all about inspiration. The building became not so much an institute housing research, but rather a symbol that people could do things differently if they made up their minds – and because someone had to actually give them permission to do things a different way. Somehow, the project was one of those magic moments where a group of people came together with a common purpose, all with the same imperative in their hearts, all working together to make it happen.

Afterword

Beyond KEEN 3 – Our Perspective

by KEEN Leadership

Since its origins in 1960, Keen Engineering has evolved extensively and traveled a long way. What began as a single office has now grown into twelve offices spanning across North America. Over the last 45 years, Keen has assembled a team of three hundred design professionals and has made impacts in many other countries. Globally, nationally and locally, Keen's work has not gone unnoticed. Perhaps the most visible and lasting effect of all is the infusion of green engineering principles into every project whenever possible. Keen has risen to the challenge of trying to create a sustainable future by continually challenging ourselves, our colleagues and the architectural/engineering community to reach new heights in green building design.

LEED®ing the Green Building Movement

Making waves in communities from afar, Keen has applied green design technologies to academics, healthcare, civic, retail, industrial, and recreation and civic centers. From small to large scale projects, we have greened virtually every facet of the building marketplace. To measure our level of actual sustainability, we often apply LEED® standards as a benchmark. As evidence of our commitment to preserving the planet, we have an extensive collection of LEED® certified projects – twelve at the close of summer 2005.

We have also enthusiastically volunteered to promote sustainability by assisting the US, CAN & World GBCs with the development of new versions of the LEED® program to broaden its practical application. Keen is proud to have been actively involved in the pilot programs for LEED® - Existing Buildings and LEED® - Commercial Interiors. We were instrumental in the first LEED® Gold in Canada, the first LEED® -EB Gold and one of the first LEED® -CI Silver projects to date. We took green design to a new level when we helped achieve the ninth ever LEED® Platinum Certified Building. As further testament to our green mission, we currently have more than thirty projects LEED® registered and awaiting certification.

Equally important to applying sustainable design principles is the transfer of knowledge and wide-scale acceptance of a green mindset. Leading the learning and training initiatives, we have served as a host for many LEED® workshops covering topics ranging from green design to becoming a LEED® Accredited Professional. Firmly committed to the professional development of the entire architecture and engineering (a/e) community, we have delivered countless presentations to our colleagues, have served on numerous committees focused on sustainability and have even delivered lectures and courses to the engineering youth at universities and colleges. We believe that promoting sustainability and "infecting" others with a green thumb is the key to setting new standards for sustainable development.

LEED® and Beyond

Beyond LEED® , there are many options for design professionals to apply sustainability in varying degrees. When designing for sustainability, one can choose to design one model facility, perhaps a Platinum LEED® certified facility... or one can choose to paint many facilities with shades of green. While a case for either approach can be made, it is clear that the impact of many green buildings will outshine that of only one – regardless of its level of sustainability. By greening facilities on a smaller scale, they will inspire more green buildings by offering a stronger sense of feasibility than even a LEED® Platinum project. Choosing to green many over creating one larger outstanding accomplishment creates a ripple effect on the rest of the community and ultimately raises their overall level of sustainability and minimum standards.

KEEN actively attempts to incorporate green into every project possible in order to maximize the positive effects of sustainability. In order to be successful in moving towards a sustainable future, we believe you must do both. A sustainable leader will continue to create exemplary facilities that inspire awe, while continuing to incorporate green into numerous facilities on a smaller scale. While we are recognized for our green accomplishments, it is most often the LEED® Certified projects that receive all the glory. At Keen, we believe there is no project too small or too remote... green does not discriminate. Creating a business-as-usual practice, we constantly try to raise the bar of sustainability with each new project. We set the minimum standards higher and we raise our sights to always achieve more.

Our Guiding Purpose

KEEN is driven by a vision to become the leading integrated engineering firm in North America with a spotlight on sustainable design. We are one team sharing one passion – of moving towards a sustainable future. To accomplish this, we offer limitless combinations of green solutions for every project. Driven by a burning desire to do the right thing, the morally responsible thing, the KEEN team actually thinks in shades of green – inside and out. Each member of the KEEN team is driven by a personal mission that is channeled into the larger organizational purpose. Not conserving resources is considered an unacceptable waste... possibly even a lack of integrity.

The power of one engineering firm is as incredible as the power of one engineer. Intensifying the current individual one tonne challenge, KEEN has issued a similar challenge to the entire a/e community - the 1,000 tonne challenge. "One engineer designing exclusively 'twice-as-good' buildings would be able, every year, to reduce [the world's] total annual greenhouse gas emissions by more than 1,000 tonnes of CO2, for a 26,000 tonne reduction of annual emissions by 2030. Multiplying annual savings by a 50-year service life for buildings, that one engineer's work saves the planet more than 1, 300, 000 tonnes."[1] The power of one is much greater than the imagination can entertain. One person truly can make a difference... a tremendous difference.

A Global Issue

In light of the prominent strengthening of the global climate change crisis, green has become a top priority worldwide. As our industrialized lifestyles are finally catching up to us, the effects of climate change

are rampant. Every nation is acknowledging the significance of the current situation and is taking serious action to become greener on a national level. The Kyoto Accord is proof that the world is not taking this challenge lightly and even those countries not in full agreement still believe that some form of serious action is necessary. At KEEN, we are working diligently to support as many communities, cities and countries that try to make their own transformation to a greener way of thinking and to do their part to combat this global issue. Although we are actively working on as many projects as possible to create a better planet, our progress is not as quick as it needs to be in order to combat climate change.

As Kevin Hydes said at the beginning of 2005, "I see a world where LEED® is not an option – it's a requirement…" Now, as we look into the future, we all agree to the immense value of a uniform standard of LEED® on a globalized level. We also realize that the standard itself needs to improve over time. As we continue to reach new heights, we must also continue to raise the bar. Soon LEED® will become the industry benchmark and then we must develop new levels beyond LEED® Platinum to challenge ourselves to stretch higher than ever before.

At our current rate of progress, change is slow from a global perspective. Assuming that we work on approximately 1,000 buildings each year, if we design each one to standards better than the average, over 50 years our work will have impacted fifty thousand buildings. At this rate it will take well over a half a century to transform North America alone. Although this number may sound impressive, we need to make a much larger impact over a shorter time period. Given the advanced stages of climate change, it is doubtful that this pace of intervention will be fast enough. Time is not on our side.

The Next Level

Realizing this truth, it became clear that we needed an accelerant... a catalyst to catapult us to the next level. As we began the lengthy process of introspection, we quickly reached a fork in the road. We could either continue down our existing path – forging our own way knowing that it might not be enough... or we could unite with a force larger than our own to help us broaden our reach and expand our practice at lightning speed. We simply could not achieve the large-scale global impact that we envision quickly enough by ourselves.

The choice was obvious. If we truly want to make a bigger difference and take action that would bring us closer to realizing our guiding vision, we need to find a partner. Not just any partner though, but a team larger than ours that has global reach, while sharing our values, goals and dreams...and naturally, our sustainable frame of mind. Rather than signifying the end of an era, this partnership is actually the same team with the same goals – just better. We aspire to use this transition to make the leap from good to great.

Building an Army of Ecological Engineers

The right partner emerged immediately – Stantec. Sharing their vision of one team with infinite solutions, KEEN elected to merge its practice with Stantec's. By creating one unified team, we can unleash the true potential of green design. As a sustainable mastermind, we can easily multiply our market transformation exponentially. This merger is definitely the case where combined efforts will easily surpass the sum of our individual efforts. Knowledge alone is nothing. It is the application of knowledge plus understanding that equals results. This new, improved green intelligence force can achieve results beyond belief – the sky is the limit.

By joining Stantec, we are positioning ourselves at KEEN to create an army of green ambassadors who will take the green revolution to the next level. As members of Stantec, we will immediately deepen our scope of services and broaden our reach. We have a moral responsibility to seize the opportunities that will help us accomplish our enduring purpose, while maintaining 100 percent integrity. Our collective mission is to create a global attitude of environmental stewardship where individuals consider every action in a sustainable context. As one world-class team, we are one step closer to creating a sustainable future.

Combining our talents provides us with the opportunity to influence the entire spectrum of the built environment. We will now be able to apply sustainable design principles to much more than buildings. We are creating the opportunity of a lifetime. We will infuse transportation, structural engineering and architecture with green technologies until we have created a team of green design professionals. The fusion of green visionaries with traditional design experts will create a larger, more effective army of Ecological Engineers.

Thank You

Throughout this book, we have shared our personal journey of discovery as we have transformed from blue to green. It has been an unforgettable experience and we are sure that the future will be equally memorable if not more so. We give special thanks to all of the people who have contributed to our arrival at this moment in time. We appreciate your unwavering support, care and encouragement. We sincerely hope that we can continue this journey together as we all seek to create a better planet for our children and our children's children.

Epilogue

Throughout my career in the engineering and architectural services community, I have had the opportunity to witness first hand the start-up, maturity and success of many firms. Few of these companies have succeeded as well as KEEN Engineering in positioning their company as a leader in their area of focus.

I have admired this small regional firm as it gradually gained momentum to become a North American leader in sustainable building design. I watched KEEN as it challenged itself to become a truly green engineering firm inside and out. The results speak for themselves. Their leadership team has propelled them to not only complete their own cultural shift to a green frame of mind but to actively support other firms in the quest for sustainability.

Today KEEN's reputation is enviable. KEEN stands for intellectual acuity – and as one of the pioneers in the green building movement, they recognized the need for sustainable design early on and put innovation into action. KEEN's ability to think creatively and turn green ideas into results have provided impressive finished products – cost-effective buildings with smaller ecological footprints. KEEN has shown outstanding leadership by leading by example.

What I also admire is their quest for continuous improvement. KEEN appears to be always on the forefront of new technologies and is constantly raising the bar. Just when clients thought that 40 percent energy savings was a milestone achievement, they delivered a new project with 50 percent savings… and then 60 percent. KEEN really does deliver on their promise to provide increasingly robust green solutions with each new project.

As this book is in its final stages, I am delighted to announce that KEEN will be merging their practice with Stantec's. Combined we will have a buildings and facilities team of over 1000 design professionals in engineering, architecture and related disciplines with more than 300 LEED® Accredited Professionals. As one team, we will be able to expand the breadth and depth of our practice and continue to raise the bar for improved technology and service for our clients.

At Stantec, we are eager to begin working with our new partners and long standing friends. We are excited about the coming opportunities to share knowledge and expertise with such a diverse range of team members and clients. I believe that this is clearly the case of two plus two equals five… possibly even more. The sum of our combined efforts will by far surpass the sum of our individual efforts.

As one world-class team we will be able to achieve more than ever before. My good friend and colleague, Kevin Hydes, will be taking on the role of leading the Stantec Building Systems Team and related sustainable development services. I know that Kevin's enthusiasm and passion will be simply contagious as we strive towards accelerating the green building marketplace transformation.

KEEN is at the forefront of a new breed of engineers in the buildings area– the Ecological Engineers. Guided by a personal mission to preserve our environment, they incorporate sustainable principles into everything they do when ever possible. This ethic is something that Stantec has always been passionate about in all of our practice areas. I am honoured to write this brief introduction to a new era and I look forward to welcoming our new colleagues.

— Tony Franceschini
President & CEO, Stantec

Appendix

Appendix I

Endnotes

CHAPTER TWO: From Blue to Green

[1] Jim Taggart, "Modern Vernacular," *Canadian Architect,* August 2002.

[2] Chris Zdeb, "Building designs that are a breath of fresh air: Windows that open featured in efficient, healthier offices," *Edmonton Journal*, June 17, 2000.

CHAPTER THREE: Being KEEN Green

[1] A.L. Marks, "Aurora Borealis".

[2] Shelley Fralic, "Form meets function in new building," *The Vancouver Sun*, April 30, 2004.

[3] Jan Niehaus, "Alberici's New Headquarters Aims For High LEED Honors," *St. Louis Construction News & Review*, January/February 2005.

[4] Stewart Brand, *How Buildings Learn*, New York: Viking, 1994.

CHAPTER FOUR: Back to the Future

[1] Kevin R. Hydes and Rosamund A. Hyde, "The 1000 Tonne Challenge," *Canadian Consulting Engineer*, June/July 2005.

[2] Tim McGinn, "Setting Fees for Profitable Green Building Projects," *ASHRAE Journal*, April 2005.

[3] Ray Cole, "Closing the Gap," *Canadian Architect*, January 2005.

CHAPTER SIX: Methodologies

[1] Also described in more detail in McLennan's *The Philosophy of Sustainable Design* and Hawken, Lovins and Lovins, *Natural Capitalism.*

[2] Comfort is determined by a host of factors including humidity, temperature, airflow and the metabolic and clothing rates of individuals.

[3] A humorous way to describe many engineers who seek to optimize their equipment without first doing reducing loads and

utilizing free resources – their solutions typically use less energy than standard approaches, but usually cost more.

[4] A parti is a word used to describe a simple sketch that can sum up a great deal of design information quickly and simply. Architects strive for clear, simple parti's on their projects to ensure that as a design progresses it does so with logic and clarity.

CHAPTER SEVEN: Low-Tech Solutions

[1] Judith Heerwagen, Ph.D., "Sustainable Design Can Be an Asset to the Bottom Line," *Environmental Design+Construction*, July 15, 2002.

[2] Natural Ventilation, in *Greening Federal Facilities: An Energy, Environmental, and Economic Resource Guide for Federal Facility Managers and Designers*, 2nd ed., edited by Alex Wilson, BuildingGreen, Inc., produced for the U.S. Department of Energy, Office of Energy Efficiency and Renewable Energy, May 2001.

[2] J.E. Braun, K.W. Montgomery, and N. Chaturvedi, "Evaluating the Performance of Building Thermal Mass Control Strategies," *International Journal of Heating, Ventilating, Air-Conditioning and Refrigeration Research*, October 2001.

[4] Werner Lang and Thomas Herzog, "Using multiple glass skins to clad buildings," *Architectural Record*, July 2000.

CHAPTER EIGHT: High-Tech Solutions

[1] "Hype Vs. Reality: New Research Findings on Underfloor Air Distribution Systems," David R. Lehrer, RA, Fred S. Bauman, PE, *Proceedings, Greenbuild 2003*, Pittsburgh PA, 2003.

[2] GreenBiz.com.

CHAPTER ELEVEN: Library Square

[1] "Interview with Moshe Safdie," *The Vancouver Sun*, April 18, 1992.

[2] *CBC Evening News*, May 25, 1995.

[3] CBC Evening News, May 25, 1995.

[4] Maria Cook, "How Vancouver's library became 'a civic meeting room'," *The Ottawa Citizen*, May 9, 2002.

AFTERWORD

[1] CCE June/July 2005 – Kevin Hydes & Dr. Rosie Hyde

Appendix II

Select Bibliography

PRIMARY SOURCES

Bauman, Fred S. “Underfloor Systems: Shedding Light on Common Myths,” *HPAC Engineering*, December 2003.

“C. K. Choi Building for the Institute of Asian Research (C. K. Choi Building),” *AIA/COTE 2000 Green Project Awards*, The American Institute of Architects, 1998-2004. <http://www.aiatopten.org/hpb/overview.cfm?ProjectID=44>.

“City of White Rock Operations Building,” *AIA/COTE 2004 Green Project Awards*, The American Institute of Architects, 1998-2004. <http://www.aiatopten.org/hpb/overview.cfm?ProjectID=288>.

Cole, Ray. “Closing the Gap,” *Canadian Architect*, January 2005.

Cook, Maria. “How Vancouver's library became ‘a civic meeting room’,” *The Ottawa Citizen*, May 9, 2002.

Dockside Green, <http://www.docksidegreen.com>.

“Double-Skin Facades and Natural Ventilation,” *High-Performance Commercial Building Facades*, Building Technologies Program, Environmental Energy Technologies Division, Ernest Orlando Lawrence Berkeley National Laboratory, University of California, Berkeley, 2002. <http://gaia.lbl.gov/hpbf/techno_c.htm>.

Haglund, Bruce and Kurt Rathmann. *Thermal Mass in Passive Solar and Energy-Conserving Buildings*, Vital Signs Curriculum Materials Project, Center for Environmental Design, University of California-Berkeley, 1996.

Hale, Tom. “Alberici's Innovative, ‘Green’ Showcase,” *Construction Digest*, July 12, 2004.

IslandWood, <http://www.islandwood.org>.

Lehrer, David R. and Fred S. Bauman. "Hype Vs. Reality: New Research Findings on Underfloor Air Distribution Systems," *Proceedings, Greenbuild 2003*, Pittsburgh PA, November 2003.

Marques, Jorge and Freda Pagani, Joanne Perdue. "Process Makes Product: The C.K. Choi Building for the Institute of Asian Research at the University of British Columbia," *ASHRAE Annual Meeting*, 1999.

McCarry, Blair T. "Innovative underfloor system." *ASHRAE Journal*, March 1998.

Laquian, E., editor. "Seeing with New Eyes", Matsuzaki Wright Architects, *Design for a New Millennium*, Vancouver: Institute of Asian Research, 1995.

Lazarus, Eve. "Victoria to make its brownfield site green: Dockside Green site will transform contaminated area to environmentally friendly community," *www.Globeandmail.com*, June 7, 2005.

Libby, Brian. "A Natural Ventilation Primer," *BetterBricks.com*, August 2003.

Nicolow, Jim. "High-Performance Daylighting: Daylight Aperture Optimization for the Southface Eco Office," *Environmental Design+Construction Magazine*, March 2004.

Suzuki, David, with Amanda McConnell. *The Sacred Balance: Rediscovering Our Place in Nature*. Vancouver: Greystone Books, 1997.

Thomas, Christopher. "Canadian colossus: Library Square, Vancouver, British Columbia, Moshe Safdie and Associates, Architect," *Architecture*, October 1995.

"Water Use Backgrounder," GreenBiz.com, a program of The National Environmental Education and Training Foundation, 2001. <http://www.greenerbuildings.com>.

Will, Gudrun. "A constructive idea," *Vancouver Courier,* June 16, 1996.

Wilson, Peter. "Interview with Moshe Safdie," *The Vancouver Sun,* April 18, 1992.

SECONDARY SOURCES

Adamson, Thelma, editor. *Folk-Tales of the Coast Salish*. New York: The American Folk-Lore Society, G.E. Stechert and Co., Agents, 1934; New York: Kraus Reprint Co., 1969.

An Architect's Guide for Sustainable Design of Office Buildings, Public Works and Government Services Canada, March 1996, revised September 1999, <http://www.pwgsc.gc.ca/realproperty/text/pubs_archguide/pubs_archguide_4-e.html>.

Besterman, Theodore. *A World Bibliography of Bibliographies*, Lausanne: Societas Bibliographica, 1966.

Centre For Interactive Research On Sustainability. Vancouver, BC: University of British Columbia, 2004.

"City of White Rock Operations Building Added to US Department of Energy High Performance Building Database", *Environmental Valuation & Cost-Benefit News*, March 23, 2005.

Collins, Jim. *Good to Great: Why Some Companies Make the Leap... and Others Don't*, New York: HarperBusiness, 2001.

"Concrete colosseum requires innovative engineering," *Canadian Consulting Engineer*; January 1994.

Curtis, Malcolm. "Radical Dockside plan wins plaudits," *Times-Colonist* (Victoria, BC), December 8, 2004.

The Living Machine®, Wastewater Technology Fact Sheet, U.S. Environmental Protection Agency, Office of Water, October 2002.

McGinn, Tim. “Setting Fees for Profitable Green Building Projects,” *ASHRAE Journal,* April 2005.

“Natural Ventilation”, Portland’s Office of Sustainable Development, <http://www.green-rated.org/resctr_tech.asp?id=6>.

O’Reilly, Dan. “Built with care”, *Heavy Construction News,* www.econstruction.ca, March 2001.

Petrie, William. *Keoeeit: The Story of the Aurora Borealis*. New York: Pergamon Press, 1963.

Procter, Don. “Unique Green Technology”. www.econstruction.ca, April 2001.

Roddick, Anita. *Take It Personally: How to Make Conscious Choices to Change the World*. Berkeley: Conari Press/HarperCollinsPublishers, 2001.

“Seminar II, The Evergreen State College,” *AIA/COTE 2005 Green Project Awards*, The American Institute of Architects, 1998-2004. <http://www.aiatopten.org/hpb/overview.cfm?ProjectID=464>.

Southface Annual Report 2004, Southface Energy Institute, Atlanta, Georgia, <http://www.southface.org>.

Appendix III

Green Reading List

The following are books recommended by the KEEN GREEN Team.

Books

A Green Vitruvius: Principles and Practice of Sustainable Architectural Design, E.A. Fitzgerald, A. McNichols, et al. London: James & James Ltd., 1999.

Agora Borealis: Engaging in Sustainable Architecture, Vivian Manasc and Cheryl Mahaffy, Edmonton, Alberta: Partners In Design Books, 2002.

Biomimicry, Janine M. Benyus, New York: Perennial, 2002.

Cradle to Cradle: Remaking the Way We Make Things, William McDonough, Michael Braungart, New York: North Point Press, 2002.

EHDD: Building Beyond the Bay, Raul A. Barreneche, New York: Edizioni Press, 2002.

Heating, Cooling, Lighting : Design Methods for Architects, Norbert Lechner, Hoboken, New Jersey: Wiley, 2nd Edition, 2000.

The HOK Guidebook to Sustainable Design, Sandra F. Mendler, William Odell, Hoboken, New Jersey: Wiley, 2000.

Low-Tech Light-Tech High-Tech: Building in the Information Age, Klaus Daniels, Elisabeth Schwaiger (translator), New York: Princeton Architectural Press, 1998.

Mechanical and Electrical Equipment for Buildings, Ben Stein, John S. Reynolds, Hoboken, New Jersey: Wiley, 9th Edition, 1999.

Natural Capitalism: Creating the Next Industrial Revolution, Paul Hawken, Amory Lovins, L. Hunter Lovins, New York: Little, Brown and Company, 1999.

The Philosophy of Sustainable Design: The Future of Architecture, Jason F. McLennan, Kansas City: Ecotone Publishing, 2004.

Residential Windows: A Guide to New Technologies and Energy Performance, Stephen Selkowitz, Dariush Arasteh, Lisa Heschong, John Carmody (Editor), New York: W. W. Norton & Company; 2nd edition, 2000

Sun, Wind, and Light: Architectural Design Strategies, G.Z. Brown and Virginia Cartwright, New York: John Wiley & Sons, 1985.

The Technology of Ecological Building: Basic Principles and Measures, Examples and Ideas, Klaus Daniels, Basel, Switzerland: Birkhäuser, 1997.

Turning Off the Heat: Why America Must Double Energy Efficiency to Save Money and Reduce Global Warming, Thomas R. Casten, Amherst, New York: Prometheus Books, 1998.

Other

"Why Build Green?", LEED-BC 'Roadmap' Workshop Summary, UBC School of Architecture, November 2003.

"Why Build Green? Ten Key Questions Answered", UBC School of Architecture, November 2003.

Periodicals

The Environmental Building News (EBN), published monthly by BuildingGreen, Inc., www.buildinggreen.com.

Environmental Design + Construction magazine, published 11 times annually by BNP Media, www.EDCmag.com.

Order Form

Ecotone Publishing

Books may be purchased by ordering on the Ecotone Website – www.ecotonedesign.com using any major credit card. Or by photocopying this order form and mailing it to us with a check made payable to Ecotone LLC. For bulk or academic orders please e-mail us at info@ecotone.com

Shipping Information

Please send all checks to Ecotone LLC

Ecotone LLC
P.O. Box 7147
Kansas City, Missouri
64113-0147

Shipping and Handling

All books are sent by regular mail. Please allow up to three weeks for delivery.

Title	Price	Quantity / Amount
The Ecological Engineer: Volume One: KEEN Engineering	$42.95 US / $52.95 CDN	
Green Dollhouse: Creating a Doll's Eye View of a Healthier World	$19.95 US / $26.95 CDN	
The Philosophy of Sustainable Design	$29.95 US / $40.95 CDN	
The Dumb Architect's Guide to Glazing Selection	$19.95 US / $24.95 CDN	
	Subtotal	
	Sales Tax (MO) add 6.92%	
	Shipping & Handling: $5.00 plus $0.50 each item	
	Total	

(Prices subject to change without notice)

Please fill out the following fields for *checks made payable to Ecotone LLC*
(for other methods of payment please visit the website listed above)

Ship To: ____________________

Address ____________________

City/State/Zip ____________________

Daytime Phone ____________________

e-mail address ____________________

About Ecotone Publishing
The Green Building Publisher

Ecotone is an independent publishing company whose mission is to educate and provide examples of restorative design to people in the building industry. In nature, an ecotone is a biologically rich transition zone between two or more dissimilar ecosystems. For architecture, it is about understanding the richness of the boundaries between the habitats of people and the environment. Ecotone – exploring the relationship between the built and natural environments.

For more information on Ecotone or to purchase other books please visit our website at:
www.ecotonedesign.com

or contact us at:
Ecotone LLC
P.O. Box 7147
Kansas City, Missouri
64113-0147

e-mail: info@ecotonedesign.com